Security of Information and Communication Networks

Security of Information and Communication Networks

Stamatios V. Kartalopoulos

IEEE Press Series on
Information & Communication
Networks Security
Stamatios Kartalopoulos, Series Editor

Celebrating 125 Years
of Engineering the Future

WILEY

A JOHN WILEY & SONS, INC., PUBLICATION

Published by John Wiley & Sons, Inc., Hoboken, New Jersey.
Published simultaneously in Canada.

For general information on our other products and services or for technical support, please contact our Customer Care Department within the United States at (800) 762-2974, outside the United States at (317) 572-3993 or fax (317) 572-4002.

Wiley also publishes its books in a variety of electronic formats. Some content that appears in print may not be available in electronic formats. For more information about Wiley products, visit our web site at www.wiley.com.

Library of Congress Cataloging-in-Publication Data is available.

ISBN: 978-0-470-29025-5

Printed in the United States of America.

10 9 8 7 6 5 4 3 2 1

To my wife Anita for her love and support

Contents

9. Network Security: Optical Systems 189

Preface

Maintaining the secrecy of information has always been a concern from the outset. As a consequence, over the centuries cryptography evolved to maintain the secrecy of data, or *confidentiality*; to assure that the transported data has not been altered, or *integrity*; if the recipient of data is the true designated one and also if the data came from the true source, or *authentication*; and that the sender cannot deny sending the data and the receiver cannot deny that the received the data, or non-*repudiation*.

In our era and within just two decades, the Internet data network spread very rapidly because the circuit-switched network was not flexible and cost-efficient enough to accommodate low-cost, new emerging data services. Thus, the legacy circuit-switched network was quickly losing its edge to an emerging data network with distributed control that met low-cost requirements even if it was not as reliable and secure.

In general, data on the Internet network is packetized and it follows a route that is not under the overall control of the network but of each node in it; that is, packets are temporarily stored in a node and hop from node to node until they reach their destination. In addition, the same packetized data crosses small and large networks, which are operated by different operators and thus they provide different quality of the service and also of data integrity.

This "connectionless" data network consists of computer-based nodes or routers and data switching works on the "store-and-forward" principle. Thus, during the time that information is stored it is accessible and vulnerable to eavesdropping, to data harvesting, and to attacks. This network structure fundamentally deviates from the circuit-switched network, which by design is data-transparent, and it meets real-time deliverability requirements.

It is a disturbing fact that the Internet network has been challenged with attacks of various forms. Destructive executable programs hiding within other programs sneak into computers where they harvest personal data, open classified files, destroy files, violate anonymity and trust, attempt fraudulent actions, clone themselves and propagate to other computers, flood the network and cause denial of service, enlist personal computers to execute programs secretly, and so on. These attacks are made from remote locations, and current incidents have placed network security on high national priority and at the forefront of research. For instance, cyber-attacks, "stealth" attacks (attacks that do not modify data or leave website traces), and silent data extraction have been on the rise, as was reported to the "Internet Security Alliance Briefing to White House Staff and Members of Congress" (by M. K. Daly, Sept. 16, 2004). The post 9/11/2001 cyber attack known as "Code Red" infected 150,000 computers in just 14 hours, and two months later the attack "NIMDA" infected 86,000 computers. Similarly, the Congressional Research Service Report to Congress (April 2004) reported that "Estimates of total world-wide losses attributable to attacks in 2003 range from $13 billion due to viruses and worms only to $226 billion for all forms of overt attacks". And in another report that was filed to the Federal Trade Commission (USA Today, 4/1/05, pp. D1), electronic heist of credit card numbers and other personal data accounts for $1/3$ of all complaints over the last three years. These reports and others have raised a serious concern by governments and industry and a new type of war has been declared, called "cyber-war," and a new type of security is in place called "cyber-security."

Until the 1980s and prior to the Internet explosion, the communications circuit-switched network also had its own share of attacks. However, the type of attacks, the number of attacks and the disaster they caused was not as extensive. Although it did not require many specialized equipment, it required networking know-how in order to tap a 2-wire pair to eavesdrop on a single conversation, as it has been captured in "spy" movies such as "Mission Impossible." It also required more specialized knowledge to mimic signaling codes with very specialized equipment, the so called "blue box," in order to mimic a source and avoid billing. However, these types of attack on the circuit-switched network have declined because the copper-based network has been replaced by fiber. Thus, fiber tapping has become more specialized and difficult, and it requires a specialist with proper equipment to extract information from signals in the fiber.

In the 1990s, the initial cellular wireless network was vulnerable to eavesdropping and calling-number mimicking. Accessing calling numbers and pin codes from the airwaves was very easy by an actor having a properly converted receiver and a laptop. Since then, cellular wireless technology and standards have been evolving by adding encryption algorithms, coding methods and more complex protocols with enhanced security and authentication procedures.

The last decade has also witnessed a formidable optical technology that has brought telecommunications to unprecedented bandwidth levels and has enabled unimaginable services. This technology uses ultra-pure silica-based fiber and laser-generated light that propagates through it and new photonic technology. With this technology and with dense wavelength division multiplexing (DWDM), many Gbps per channel or several Tbps in a single fiber are reality. Although this technology is more complex than its predecessors, nevertheless, it too is vulnerable to bad actors with proper know-how and sophisticated tools. Therefore, security of the network is extremely important, regardless of the medium and of the difficulty of the cryptographic system. In fact, it has been recognized that:

> *The best cryptography does NOT warranty secrecy if computers are not immune to attacks; and the most immune computer does NOT warranty privacy if the network is NOT secure.*

Therefore, a concerted effort is needed so that cryptographic systems are unbreakable, computer-based nodes are attack resilient, and the network is secure, self-defensive, and counter-attacking with countermeasure strategies. This book examines the security of all three: the cryptographic system, the computer and its attacks, and the network.

From a cryptographic viewpoint, various provable cryptosystems and public-key protocols have been developed to secure the privacy of documents and to protect them from electronic copying, cloning, and destroying, and to also deter unauthorized entry into the network.

AES is a symmetric cipher and it is based on the "Rijndael" algorithm (invented by V. Rijmen and J. Daemen) and it specifies three approved key lengths, 128, 192 and 256 bits. The AES is a federal government approved encryption algorithm defined in Federal Information Procedure Standard (FIPS) no. 197 (2001). NSA has approved the 128-bit AES for use up to SECRET level documents, and the 192-bit AES for up to TOP SECRET level.

A hash function takes an arbitrary amount of input and produces an output of fixed size, known as *digest*. The output is always the same for the same input and it is impossible to determine the input given the output.

The public-key cryptography, introduced by W. Diffie and M. Hellman in 1976 [3] uses a pair of keys, a publicly available key and a private key. A well-known public-key cryptosystem is the Elliptic Curve Cryptography (ECC). The U.S. National Security Agency (NSA) has already recommended a set of advanced cryptography algorithms known as Suite B for securing U.S. government sensitive and unclassified communications. The Elliptic Curve Menezes-Qu-Vanstone (ECMQV) and the Elliptic Curve Diffie-Hellman (ECDH) for key agreement, the Elliptic Curve Digital Signature Algorithm (ECDSA) for authentication, the Advanced Encryption Standard (AES) for data encryption and the Secure Hash Algorithm (SHA) for hashing are public key protocols included in Suite B. However, no matter how smart or difficult a cryptographic algorithm is, sooner or later someone will outsmart it and break it. In fact, a research effort known as steganalysis is doing just that: using sophisticated methods with advanced filtering techniques. Steganalysis examines encrypted methods and retrieves hidden messages in them. Thus, it is no surprise that researchers seek refuge in unconventional methods, such as quantum mechanics, in search of the "holy grail" of cryptography.

The development of cryptographic protocols for key establishment and authentication led to a reverse activity in an effort to evaluate their robustness, *cryptographic protocol analysis*, or *cryptanalysis*. Two methods have been used: the *computer security approach* and the *computational complexity approach*. The computer security approach focuses on automated machine specification and analysis, such as model checking and theorem proving. The computational complexity approach follows deductive reasoning using a proven approach for breaking a protocol to another protocol that is under analysis. However, cryptanalysis by cryptographic protocol developers as a reverse activity may have potential flaws because it does not realistically emulate the tools the adversary has and the approach that is used. As a result, it is typical that an open challenge to code-breakers is announced with a reward. Thus, cryptanalysis does not prove that cryptography is robust but that cryptography is not, when the code is broken.

In this book:

- Chapter 1 provides an introduction to the subject of cryptography (from ancient to modern era), computer and network security.

- Chapter 2 provides a simplified mathematical foundation that is necessary in the understanding of advanced cryptography.

- Chapter 3 presents many well-known symmetric and asymmetric key cryptographic systems, including elliptic curve cryptography.

- Chapter 4 describes other well-known cryptographic algorithms such as Shamir's. Diffie-Hellman, Digital signatures, trusted third parties and key escrow systems.

- Chapter 5 provides a simplified description of chaotic systems and we outline some new methods.

- Chapter 6 distinguishes and classifies different communications layers of security that pertains to the application layer, the MAC/Control and the physical layer (PHY).

- Chapters 7 through 9 provide a description of how security is applied in various communication networks with wireless access, wired access and optical access. They also describe vulnerabilities of the media, potential attacks, methods for detecting attacks and countermeasure strategies.

- Chapter 10 introduces the advanced topic of Quantum Cryptography, identifies current vulnerabilities, and describes the most current research topic on teleportation providing a trivialized example for non-research readers.

- Chapter 11 provides an introduction to security aspects and issues related to the next generation optical networks, such as the next generation SDH, Fiber-to-the Premises, IPvN, and so on.

- Chapter 12 provides a short description of Biometrics, how biometric technologies are used in the communications network, and their security issues.

The purpose of this book is educational and informative and also to challenge the intellect of the reader; the material is derived from research and standards, and the intention is not to provide comprehensive guidelines for design. Although it is addressed to a general audience, it is suitable to undergraduate studies, to graduate studies, to researchers, and to all those interested in a general education in the security of information and communication networks.

Happy reading.

Stamatios V. Kartalopoulos, PhD

Acknowledgments

To my wife Anita, son Bill and daughter Stephanie for consistent patience and encouragement. To my publishers and staff for cooperation, enthusiasm, and project management. To the anonymous reviewers for useful comments and constructive criticism. And to all those who worked diligently on the production of this book.

About the Author

Stamatios V. Kartalopoulos, PhD, is the Williams Professor in Telecommunications Networking with the University of Oklahoma, EE/ECE/TCOM graduate program. His research emphasis is on optical communication networks (including access PON/FTTH and FSO), on optical technology (including optical metamaterials), and on optical network security including quantum cryptography and quantum key distribution protocols, chaotic processes and biometrics.

Prior to academia, he was with Bell Laboratories where he defined, led and managed research and development teams in the areas of DWDM networks, SONET/SDH and ATM, Cross-connects, Switching, Transmission and Access systems, for which he received the President's Award and many awards of Excellence.

He holds nineteen patents in communications networks, has authored eight reference textbooks and more than hundred seventy scientific papers, and he has contributed chapters to several books.

He has been an IEEE and a Lucent Technologies Distinguished Lecturer, and has lectured world-wide at universities, NASA and conferences. He has been keynote and plenary speaker at major international conferences, has moderated executive forums, has been a panelist of interdisciplinary panels, and has organized symposia, workshops and sessions at major international communications conferences.

Dr Kartalopoulos is an IEEE Fellow, chair and founder of the IEEE ComSoc Communications & Information Security Technical Committee (CISTC) and past chair of ComSoc SPCE and of Emerging Technologies Technical Committees, member at large of IEEE Biometrics Council (previously NTDC), editor-in-chief of IEEE Press and Area-editor of IEEE Communications Magazine/Optical Communications, member of IEEE PSPB, VP of IEEE Neural Networks Council (now Computational Intelligence Society) and member of the IEEE Experts-now program.

Chapter 1

Introduction

1.1 A HISTORICAL PERSPECTIVE OF INFORMATION AND NETWORK SECURITY

1.1.1 Hidden Messages

Delivering messages in secrecy has been a serious concern since antiquity. Messages that conveyed personal, business, or state affairs were very critical for the well-being of a person, business or country, and as history has shown even in more recent times, the outcome of a war depended on the prompt and safe delivery of a critical message. The players involved in the transport of a secret message are the author and rightful sender, the courier or the transporting medium, the authorized receiver and the interceptor. Because the sender of the secret message was aware that there are those other than the authorized recipient who would attempt to gain knowledge of the content of the secret message, the sender used a coding method to encrypt the message and assure secrecy. The courier was initially a trusted person who at risk of life had to deliver the message to the authorized recipient. The interceptor, depending on sophistication and opportunistic factors, had several choices: attack and capture the message; attack and destroy the message; acquire knowledge of the message content but do not alter it; get hold of the message, alter it and send it to the recipient. The authorized recipient of the message should be able to decode the message, verify the authenticity of the received message and also detect if the message was intercepted and altered.

The lessons learned over time forced senders to use more and more complex cryptographic methods to outsmart sophisticated and knowledgeable attackers.

Security of Information and Communication Networks, by Stamatios V. Kartalopoulos
Copyright © 2009 Institute of Electrical and Electronics Engineers

Appreciating the art of cryptography, its sophistication, and how it evolved to its current status requires a review of some characteristic examples.

Ancient Mesopotamians wrote a private message in cuneiform script on a fresh clay tablet, which was exposed to the sun to dry. This tablet was then enclosed in a clay envelope on which the addressee's name was written (Figure 1.1). When the envelope was dry, it was dispatched with a trusted messenger. If this was intercepted, the clay envelope had to be broken, revealing to the recipient that the message was compromised. In other cultures, clay was substituted by papyrus, tree bark (such as white birch), carta-pergamena or parchment (processed skin of baby lamb or goat), or by fabric or paper on which text was written with a stylus and ink, and then rolled or folded and sealed with Spanish wax on which a symbol was impressed using either a signet ring or a stamping tool.

In addition to fabric and paper, the ancient Chinese had a method of hiding written messages hidden in cakes (known as *moon cake*). This cake was given to a trusted agent who passed it through the unsuspected guards of the gate. Similarly, the Egyptians used their own secret methods as is discerned in hieroglyphs, and so did the ancient Greeks as is discerned in ancient texts. It's worth to outline the following three examples because they have some similarities with modern encryption methods.

In all cultures, the trusted courier has been the most popular method of sending a secret short message, which was memorized so that there was no physical evidence of the message. A notable example is that of the soldier Philippides who ran from Marathon to Athens, a distance of about 40 km, to deliver to the Athenians the message that the battle at Marathon was *won*, a decisive victory for spreading civi-

Clay envelope with address markings on it

Message on clay tablet

Figure 1.1.   Clay tablets enclosed in clay envelopes assured secrecy and authenticity of the message.

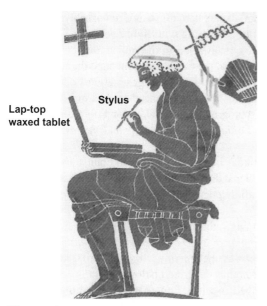

Figure 1.2.   Hinged wooden tablets on which wax was spread was the scratch pad of antiquity. Scraping the wax, writing a secret message and spreading the wax on top of it was a form of cryptography of yester day.

lization and democracy; the modern *Marathon race* commemorates this victory and Philippides' accomplishment.

   The Athenian Demeratus used a different method to transport a written message through the enemy ranks, which gave birth to the modern terms "*cryptography*" (from *crypto* and *graphe* or "hidden message") and also "*steganography*" (from *steganos* and *graphe* or "sealed message"). He used waxed tablets (a popular writing medium like a paper scratch pad is today) that were hinged like a laptop (Figure 1.2). He scraped off the wax, wrote the message on the wooden surface with a carbon stick (a pencil) and then spread the wax over it, on which he wrote with a stylus an unclassified message. According to Herodotus, Demaratus' method worked very effectively and efficiently [1].

1.1.2   Encoded Messages

For long distances and for rapidly transmitting messages, the Greeks understood this can be done only with light. To accomplish this, small towers were built on top of hills and mountains forming a network. From these towers, optical messages were transmitted quickly, reaching far destinations. This method is discerned in Aeschylus' play *Agamemnon*, the Greek campaign general in the Trojan War (ca. 12th c. BCE) [2], and currently the method has been dubbed the *Agamemnon's Link*. According to Aeschylus, it took only few hours for a message to arrive at Argos from Troy, a distance more than 600 kilometers; 3000 years ago, this was a remarkably short time [3]. Similarly, Herodotus wrote of communicating with light over long distances at the battle of Thermopylae (captured by Hollywood in the movie *300*).

What is not known is if and how optical messages were encoded then. However, historians do know that by about 350 BCE, optical messages were encoded according to a method developed by the military scientist Aeneas Tacitos of Stymphalos. According to the Greek historian, cryptographer and navigator Polybius (203–120 BCE), *communicating with encrypted light messages became the greatest service in war* [2].

It is believed that the encryption key changed from hour to hour (a method now known as the *cryptoperiod*) using a clepsydra, a water clock made with a leaky jug which had markings on the interior calibrated so that the water level in it determined the cipher key to be used.

Polybius, too, wrote about encoding light messages. Although his method was initially invented by *Cleoxenos The Engineer* and *Demokleitus The Inventor*, Polybius perfected it, and it is currently known as the *Polybius square*. This method is of current interest to cryptography, and is thus described below.

The alphabet letters were arranged in a 5×5 matrix; this example uses, the Latin alphabet and therefore since there are 26 letters, two least used letters (Y and Z) are placed in the same matrix element. Each row of the matrix is written in a tablet, letters on the tablet are numbered left to right from one to five, and each tablet is numbered from one to five (Table 1.1).

Both the transmitting and the receiving site had five lit torches. When the transmitting site wanted to send a message, the guards raised two torches, and the receiving site in response raised two torches as well, which were subsequently lowered; this established a *"request to send"* and the *"acknowledgment"* steps of modern communications protocols.

Now, to transmit the message VICTORY, the message was written on a tablet and each letter was encoded by writing the tablet number where the letter is, followed by the letter number on the tablet; the word VICTORY was then transformed to a sequence of two-digit numbers: 52, 24, 13, 45, 35, 41 and 55. Now, to send the first letter, five torches were raised and lowered followed by two torches; to make up the number 52 (for letter V). Then, two torches were raised and lowered, followed by four torches to make up the number 24 (for letter I), and so on.

Table 1.1. Arranging the alphabet in a matrix and on numbered tablets

	1	2	3	4	5
Tablet #1	A	B	C	D	E
Tablet #2	F	G	H	I	J
Tablet #3	K	L	M	N	O
Tablet #4	P	Q	R	S	T
Tablet #5	U	V	W	X	Y/Z

As trivial as this may seem, because of the distance between two towers and in order to remove errors based on human factors, Polybius provides training guidelines, optimum distance between torches, dimensions of relay towers, "viewing tubes" and screens, and more. What he did not publicize, for obvious reasons, is whether the letters of the alphabet were arranged on the tablets in an linear order as in the preceding example or if they were arranged in a random order, in which case all stations on the transmitting path should have copies of the same tablets, which also could change periodically to assure secrecy of the code (Table 1.2).

In addition to light as means of transmitting messages, Spartan ambassadors used another sophisticated method. Wrapping a ribbon helicoidally around a baton or staff of specific diameter that the Spartans called "skytale" (this was carried by all ambassadors), and then writing a message alongside would produce an unintelligible message on the unraveled ribbon; the method was recently used in a Hollywood movie (Figure 1.3). The message could only be read if the recipient had a baton of the same diameter with the original and if the ribbon would be wrapped around the baton in the same helicoidal sense and direction. Figure 1.4 illustrates the same phrase but wrapped in the opposite sense yielding an unintelligible message.

During the Roman era, Julius Caesar (100–44 BCE) proposed an encryption method, hence known as *Caesar's Cipher*. This method shifted the letter in the message to another letter of the alphabet in a linear manner. For example, if a letter in a message is replaced by the next letter in the alphabet, the word *CAESAR* becomes *DBFTBS*, and if by the one after the next, then it becomes *ECGUCT*. Notice however that with this algorithm the frequency of occurrence of letters (such as A in Caesar) remains, although encoded to another letter; the frequency of occurrence of letters in words in a text reveals a vulnerability that modern crypto-analysts use to break the cryptographic algorithm and find the cipher key. The Caesar's algorithm is considered trivial by modern cryptographers, who extended it to a more general encryption algorithm, by which each letter of a text is replaced by another letter of the alphabet according to a specified random shifting algorithm. This cryptographic method is currently known as *Shift Cipher* and it uses randomized arithmetic based on modulom operations.

In recent times and until WWII, German intelligence used an encryption method known as the *enigma*, based on a modified typewriter, the *enigma typewriter*, which

Table 1.2. Arranging the alphabet in random order provided additional message security

	1	2	3	4	5
Tablet #1	A	E/Z	I	G	M
Tablet #2	J	B	Q	U	S
Tablet #3	C	V	D	O	X
Tablet #4	T	Y	N	W	P
Tablet #5	H	L	F	R	K

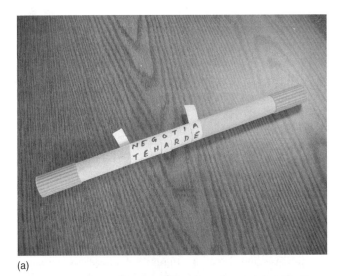

(a)

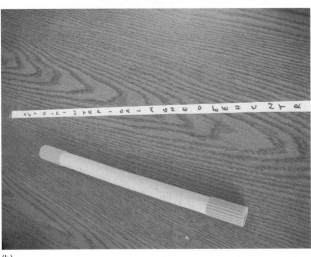

(b)

Figure 1.3.   (a) A staff and a ribbon wrapped around it provided means of encoding a secret message such as "NEGOTIATE HARDER HELP IS ON THE WAY" (b) The unwrapped ribbon was unintelligible: AESGIDINTRIOALMGHEOEEHCNTR.

encrypted messages as they were typewritten. The enigma typewriter consisted of three alphabets that rotated separately after an alphabet key was depressed, thus yielding a combination of $26 \times 26 \times 26 = 17{,}576$ alphabets [4, 5]; these were reflected by a "mirror" causing three more transpositions. Thus, as the operator typed each letter of the message on the keyboard, the enigma typewriter scrambled it to another, producing an unintelligible message. The exact relationship of the three alphabets as they rotated on the drums established the "cryptographic key." Conversely, by

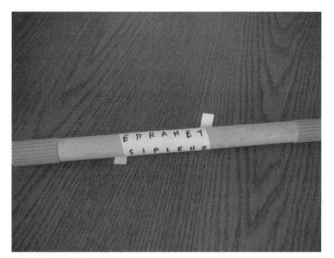

Figure 1.4.   The ribbon of Figure 1.3 wrapped in the opposite sense also yields an unintelligible phrase.

typing the scrambled message on the enigma typewriter with the same key, the original message was recovered. Clearly, this method required two exact replicas of the enigma typewriters to cipher and decipher a message and therefore transporting the machines and the keys was done in utmost secrecy because the other side was very eager to know how the machine worked and what the key was. A WW II German Enigma machine is kept in the United States National Cryptologic Museum (in the collection of artifacts related to cryptology), located adjacent to the National Security Agency (NSA) Headquarters, Fort George G. Meade, Maryland.

With the evolution of radio transmission and particularly during World War II, message encoding gained particular importance because electromagnetic waves would reach both friendly and foe antennas. As a result, cryptography entered the realm of science (modulation methods) and mathematics (statistics, probability, and number theory) and several encryption algorithms were developed to provide authentication, no repudiation, data integrity, and confidentiality. Such algorithms required two fundamental elements: a unique *cipher key* that ciphers and deciphers a message, and a unique *key distribution method* so that no one can successfully intercept it.

In parallel to this, another effort of equal importance has been on the way: intercept the transport layer, copy the secret message (or cipher text) and try from this to figure out the secret key (that is, perform an automatic analysis of the cryptographic algorithm), or interrupt the key distribution process to disable secure communication. This effort can be used to test the hardness of the cryptographic method, the robustness of the distribution mechanism and also identify vulnerabilities; this is known as *cryptanalysis*. On the negative side, cryptanalysis can also be used by malicious attackers. It is typical that new encryption methods are internationally challenged at a prize for unbreakability or breakability of the cipher text and tolerance of the key distribution method.

1.2 MODERN CRYPTOGRAPHY, WATERMARKING, STEGANOGRAPHY, ESCROW AND CRYPTANALYSIS

1.2.1 Cryptography

Cryptography is a method that, based on an algorithm or some other method, transforms plain text (data) into unintelligible or undetectable *cipher text* to all but the authorized recipient who possesses special knowledge to convert the cipher text back to the original plain text. Based on this, if the cipher text is viewed by an unauthorized party, an eavesdropper or malicious attacker, the original text cannot be read from it; thus, unintelligible means that the plain text has been transformed to an alpha-numeric or binary string that makes no sense, or that the cipher text may seem intelligible but it is not the original plain text. For example, consider the plaintext "TRANSFER_ACCOUNT_TOMORROW_AM". Using a cryptographic algorithm, this is transformed to the unintelligible string "SIGPZFKV_BCIRQQT_QNPRQSX_GP," or using double meaning linguistic to the undetectable cipher text "BEES GO FOR HONEY AT SUNRISE," or some other intelligible phrase that triggers some specific action without explicitly revealing to a third party what the action is. The latter is widely used in many cases particularly in the classified section of newspapers, over radio waves and TV; only the knowledgeable receiver understands the exact meaning of the message.

In digital communication systems, information is transmitted in binary form; alphanumeric, math and other symbols in a text are each converted in eight-bit bytes and transmitted as a string of ones and zeros. However, prior to transmitting this binary string, the string is scrambled with a cipher key that preferably is as long as the string. As an example, if the message in binary notation is 10001101|00001101|01001101| and the established cipher key is 01010100|11110110|00011011|, where the vertical bar | delineates each byte, then the cipher text is obtained by bit modulo-2 or Exclusive OR (XOR) logic operation yielding the cipher text 11011001|11111011 |01010111|. If C_k is the cipher-key and T_x is the plain text then the cipher text C_T is $C_T = C_k \otimes T_x$. The XOR function is very convenient because at the receiver, the original plain text is recovered from $T_x = C_T \otimes C_k$ as a result of the XOR identity (if $A = B \otimes C$ then $B = A \otimes C$ and $C = A \otimes B$). This is known as *decrypting* or *deciphering*.

It is evident that deciphering requires that the cipher text has not errored bits as a result of noise, attenuation and pulse shape deformations. Similarly, it is important that the cipher key is bit synchronized with the cipher text.

In the example above, when the message is represented by a block of bytes, the cipher key is known as *block cipher*. Certain encryption algorithms use a long key to create a cryptographically strong *keystream*. It is this keystream that is exclusive-ORed with the plain text; this algorithm is known as *stream cipher*. Cipher keys may be *permanent*, they may change periodically (in which case a *cipherperiod* is defined), or they may change for each message (known as *one-time pad keys* or *ephemeral keys*).

1.2.1.1 Symmetric and Asymmetric Keys

There are two fundamental cryptographic methods in ciphering/deciphering. One uses the same key at both ends of the channel and the other different keys at each end.

- When the cipher key is the same at both ends of the cryptographic channel, then cryptography is referred to as *symmetric cryptography* and the key as *symmetric key*. To generate the symmetric key another symmetric key may be needed first, the *seed key*, which is used to calculate the final cipher key, hence known as the *key encrypting key*; this process is known as *key wrapping*. In applications where the text is binary serial (in data communications), the cipher key operates on the serial bit-stream a bit or a byte at a time using XOR. Such cryptographic methods are also known as *stream ciphers*. The stream cipher text is decoded serially by applying the same XOR logic operation with the same key (as in the aforementioned example). When the key is generated by a pseudo-random generator, it is called a *key-stream generator*. One of the issues with symmetric cryptography is the *key distribution*. That is, the transportation or communication of the key from the sender of the cipher text to the rightful recipient of it. If the cipher key is intercepted and copied by an intruder during its transportation, the value of the cipher text becomes meaningless. However, symmetric cryptography may become stronger if the symmetric key changes often and randomly according to a secretly timed algorithm.

- When the cipher key is not the same at either end of the cryptographic channel, then cryptography is referred to as *asymmetric cryptography*. Based on this, the encoding end uses one key that keeps it secret, and the decoding end another key that also keeps it secret; the two keys are mathematically interrelated. Thus, even after intercepting the cipher text, it is very difficult to decipher it or identify the decoding key. In some asymmetric cryptographic systems, a publicly transported key is needed to calculate the decoding key, known as a *public key* and hence *public key cryptography*; in this case, the public key is transported over a *public channel* (such as wireless). Intercepting the public key does not help in decoding because the mathematical algorithm that produces the deciphering key is either secret or too difficult. Thus, the asymmetric key method has a lot of appeal and it is the method that has inspired some more complex cryptographic methods applicable to the Internet and others. To summarize, public key cryptography uses a pair of two key ciphers, a public and a private. Plain-text encrypted with the public key can only be decrypted with the associated private key. Public key cryptography is also used in conjunction with one-way hash functions to produce digital signatures. According to this, messages signed with the private key can be verified with the public key [13, 14].

1.2.1.2 Hash Functions

Several algorithms map a bit string (or plain text) of arbitrary length to a bit string of fixed length, known as *message digest*, according to an approved function known as *one-way encryption* or *hash function*. The result of applying the hash function on the bit string is the *hash value* or a *digital fingerprint* of a file's contents. Thus, hash functions provide a measure of the integrity of a message. Hash functions satisfy the property that it is computationally infeasible to map any input string length to a pre-specified output bit string, and also that it is computationally infeasible for two distinct input strings to map onto the same output pre-specified string. Thus, the output is always the same for the same input.

There are two primary hash functions in use, the message digest 5 (MD5) and the secure hash algorithm-1 (SHA-1).

- MD5 was developed by RSA labs and SHA-1 by the National Security Agency (NSA).
- SHA-1 supports messages up to 2^{64} bits at the input and it produces a 160-bit digest [6–8].

The *Secure Hash Algorithm* (SHA) is a public key protocol included in Suite B, consistent with the National Institute of Standards and Technology (NIST) publications. Yet, vulnerabilities or insecurities are not absent from these algorithms [6] and recently it was announced that the SHA-1 algorithm was broken by a Chinese research team (March 1, 2005, *The Australian*). In cryptosystem language, the public and private keys, symmetric keys, block ciphers, and hash functions are known as *cryptographic primitives*.

1.2.1.3 Security Services

The information security services that are performed in cryptography are *authentication* (of source and destination), *authorization*, data *integrity, data privacy*, message *confidentiality*, and *non-repudiation*.

- *Authentication* is a process that verifies the identity of source and/or destination, verifies that the received message was sent by the rightful sending entity, and that the received message has not been altered during transmission [9–11]. Authentication is accomplished with cryptographic checksums known as the *authentication code* calculated according to an approved cryptographic algorithm; the authentication code is also known as *message authentication code*. A message authentication code is a one-way hash function computed from a message and a secret key. It is difficult to forge without knowing the secret key.
- *Authorization* is a process that grants access privileges to an entity such as the intended destination or to a third party. Authorization to the destination is granted after the destination authentication. Access authorization to a third party is granted only after an official and justified request to access a message

and perhaps modify the message. Access is under the control of an *access authority* (or function), which is responsible for monitoring and granting privileges to other authorities that request access.

- *Data integrity* pertains to identifying possible unauthorized alterations in the transported information. In communications, one of the mechanisms to secure document or data integrity is digital signatures and steganography; that is a non-removable signature that authenticates the original text or invisible text superimposed to the original text which only the rightful recipient knows how to remove. Thus, when part or all of the text is altered by an unauthorized entity, the signature or the steganogram has been altered as well.

- *Data privacy* secures the authorship of data, such as for example the ownership and copyright of a text, picture, or movie. This is accomplished by superimposing a visible text or a symbol (*A*) over the original data (*B*) such that the two (*A* + *B*) become inseparable using a cryptographic key; this is known as *watermarking*. Altering watermarked data also alters the original data. Thus, watermarking does not make a text unintelligible and undetectable, but it secures the ownership of it. A common application of watermarking is the word *draft* or *classified* across the page of text, or the name of the photographer on a picture.

- *Message confidentiality* is the service that warranties that information during its transport from source to destination will not be disclosed to one or more unauthorized parties. Cryptography with strong encoding is the mechanism that ensures message confidentiality. The network also provides an *accountability* function that monitors and ensures that actions of any entity on the network security can be traced back to it.

- *Non-repudiation* provides proof of the integrity and of the origin of a message to a third party. Non-repudiation prevents an entity from denying sourcing or receiving a message. For example, a signed message provides undeniable proof that the message was sourced by the transmitting entity and a coded time stamp that the message was received by the rightful owner. The digital signature is calculated with a secret key. In similar applications, the equipment signature ID of the sourcing computer is embedded on data prior to transmitting, and the signature ID of the receiving computer is embedded as soon as it is received. The equipment signature ID is a unique code that identifies the particular end device; it is encoded and it cannot be duplicated.

1.2.2 Watermarking

Digital watermarking is a cryptographic method that safeguards the copyright and the ownership of data and its privacy. Watermarking is not designed to embed a hidden message like steganography does, but to fuse a notice (*A*) with a target file (*b*) that contains text, image (photo, video, etc.) or audio, so that an unauthorized user cannot extract from the fused file (*A* + *B*) the target file. Based on this, stock

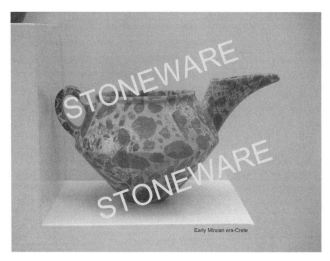

Early Minoan era-Crete

Figure 1.5. An image fused with a label produces a watermarked image, which can be separated with a key that only the owner of the image holds.

photography agencies that display their images on the web, watermark those images with their logo or agency name; when an end user purchases the rights to use, then the agency removes the watermark (Figure 1.5). The watermark, at the pixel level, has typically lower amplitude than the target file and it appears semitransparent when superimposed with text or image. In a different application, watermarking has been used on paper money to warranty authenticity. Commercial software watermarks text or image.

1.2.3 Steganography

A form of cryptography known as *steganography* is a technology that embeds a secret text (*A*) within a non-secret text (*B*) the *carrier* to form a composite text (*A* + *B*) known as *steganogram*. In this case, the steganogram looks like the intelligible and detectable text *B*, but embedded text *A*, is only detectable or intelligible to the rightful and authorized recipient who holds the proper *steganographic key*; to the naked eye, the hidden text *A* is invisible.

Steganography takes advantage of quantization noise after digitizing an analog signal (picture, voice or text). Thus, if the steganogram (*A* + *B*) has the same statistical characteristics of the carrier signal *B* and the *steganographic key* is not known, then it is very difficult to extract signal *A* from the steganogram. We can parallel steganography with traditional telephony that encodes analog signals using μ-law coders; because some small bandwidth is needed for signaling, the least significant bit of certain time slots is superimposed by signaling bits, a method known as *bit robbing*; this is a well-tested and well-used method and it does not affect the quality of voice signal when it is heard by the callee. Thus, bit robbing may be considered the first form of steganography in telecommunications. One of the applications in steganography uses a digitized color picture in which some bits at random (according to a secret key) have been altered to host bits of a digitized secret text. To retrieve the secret message in it, one must know the algorithm and find the embedded bits in it.

1.2.4 Escrow

In cryptography, the term *key escrow system* or *escrow* means that the two components that comprise a cryptographic key are entrusted to two key holders called *escrow agents*. The escrow agents provide the components of the key to a *grantee* entity only upon fulfillment of pre-specified conditions. The grantee entity reconstructs a unique key from the two components and generates the session key, which is then used to decrypt the cipher text [12].

An overall end-to-end cryptographic process consists of the transmitter or source with an algorithm that generates a secret key A, the receiver or destination with another algorithm that generates a secret key B, and the link or medium between the two ends which may be attacked; the attacker has means for capturing or discovering secret keys (Figure 1.6). The major functions at the two ends are encapsulated as:

- An algorithm at the transmitter that generates a key (or cipher key); this key may consist of more than one component.

- A secure mechanism for transporting the key to the receiver; the transported key may or may not be the same with that at the source. The method by which the cipher key is exchanged and agreed upon by both sender and receiver is known as *key establishment*.

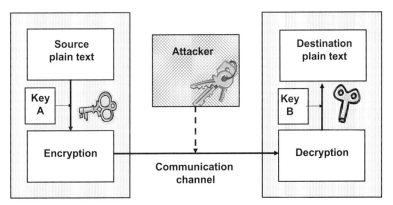

Figure 1.6. At both ends of a communications channel the plain text is ciphered using a secret key and is deciphered to plain text using the same or another cipher key. Bad actors evaluate vulnerabilities of the channel and cryptographic systems to attack.

- A secure process and a key *registration authority* that registers the agreed upon key in a *key depository* or *key management archive.*

- A secure mechanism at the transmitter that transforms the original text (or *plain text*) to an encrypted text (or *cipher text*). This process is called *encoding* or *ciphering.*

- A secure medium for transporting the cipher text to the rightful recipient.

- A protocol at the receiver that authenticates the source and one that authenticates the destination. In other words, a *public key certificate* that uniquely identifies an entity, contains the entity's public key, uniquely binds the public key with the entity, and a trusted *certificate authority* that digitally signs it, as part of the X.509 protocol (or ISO Authentication framework).

- An algorithm that transforms the received cipher text back to its original form, the plain text, using the decipher key. This process is called *decoding* or *deciphering.*

1.2.5 Cryptanalysis

If cryptography tries to cipher a text and make it unintelligible to an unauthorized agent, cryptanalysis tries to break cryptography and discover the cipher key or read the unintelligible cipher text using various methods that include mathematics, social engineering, deception and other useful clues that help to break the code.

In its pure form, cryptanalysis does not necessarily know the cipher key or the cryptographic algorithm and therefore it is much more difficult than cryptography. Examples include the decipherment of ancient scripts that were unreadable for millennia (such as, the Mesopotamian cuneiform, and the Mycenaean Linear B scripts, the Mayan glyphs and the Egyptian hieroglyphs), as well as some that are still unbreakable (such as the Harappan and the Minoan Linear A scripts). The term cryptanalysis is a compound word from Greek *kryptós* (hidden) and *analysis* (to untie or to *analyze*).

Cryptanalysis is also used to test the hardness of cryptography. The faster a cipher text is deciphered with cryptanalysis, the softer the cryptographic algorithm is. Therefore, new cryptographic algorithms challenge cryptanalysts and offer a reward or a prize if and when they break the secret key. On the negative side, bad actors also try to break the secret key for their own profit in any way they can, using cryptanalysis and also other illegal and unethical means (including social engineering tactics). Hollywood has captured this in many spy movies.

Malevolent cryptanalysts also use methods for facilitating and expediting the code breaking. Among them are capturing a plain text and its corresponding cipher text from which they determine the cryptographic system and algorithm; this is extremely valuable information because they can decipher future cryptograms much more easily, even if the cipher key has changed. In certain cases, bad actors impersonate a third party and transmit a short plain text to the source, which encrypts it; capturing the encrypted message will provide the desirable result to the bad actor.

In a different scenario, the attacker may capture the same plain text that has been ciphered with two different keys. Although the keys are not known, their interrelationship is found as well as the workings of the cryptosystem.

Currently, cryptanalysis is classified as either linear or differential.

- *Linear cryptanalysis* is based on probabilities and it takes advantage of the statistical occurrences in the cipher text. The premise is that the cipher breaker already has some statistical information of known plain texts and their corresponding cipher texts. Thus, if the cipher process is considered to be a black box, its input the plain text and its output the cipher text, the cipher breaker tries to discern the general workings of the black box.

- *Differential cryptanalysis* exploits the statistical behavior of plain text differences and the differences in corresponding cipher texts. That is, this class of cryptanalysis examines the difference ΔX of plain texts X_i, and the difference ΔY of their corresponding cipher texts Y_i at the output of the black box. Differential cryptanalysis assumes that the attacker has knowledge of or is able to select plain texts and also to observe their corresponding cipher texts from which ΔX and ΔY are calculated.

In conclusion, with modern cryptography there is also modern cryptanalysis, which is based in problem or puzzle solving techniques using computers. Because a cryptosystem may be based on symmetric or asymmetric keys and on different mathematical approaches, cryptanalysis may use many computers to collectively and in parallel try to solve the cipher problem. With fast execution time, computers have been successful and have broken several codes.

1.3 NETWORK SECURITY

Up to this point, we defined vulnerabilities associated with computer-based nodes and end-terminals, which have attracted a number of different attack types. In traditional synchronous communications, accessing the loop of a circuit-switched network required moderate networking know-how to tap a two-wire pair to eavesdrop on a conversation, and more know-how to mimic signaling codes with the so called "blue box," and establish end-to-end connectivity without being billed. Despite these attacks, the synchronous network was not the subject of attacks as the data networks are. The reason is that data did flow continuously without being buffered or stored, as a characteristic of the circuit switched method. In addition, demultiplexing time-slots at the core network required specialized equipment and substantial know how of complex protocols and time-slot assignments. Thus, the security of the synchronous network was not challenged with virus attacks by outside bad actors other than eavesdropping on individual conversations at risk of been caught. Virus and other malicious soft attacks are associated with Internet and computer communication networks for which many "cyber-security" reports have been drafted for governments and enterprise [13–16] in an attempt to develop counter cyber-attack strategies and thwart the humongous destructive attempts and permit to trace back and discover the offending terminal and its user. Unfortunately, such attempts can be made remotely (from another country) by high school and college students who don't have specialized equipment, just personal computer, and who are not seeking profit but to demonstrate intelligence!

In general, information assurance and security aims to ensure a level of trust to the client and by the client commensurate with client expectations. Such expectations include information or data protection during its creation, use, transformation, storage, and transport. In addition, the expectation is that data are not retained at layer boundaries of the reference model (such as the ISO, ATM, TCP/IP), and at the transport layer. In a computation environment, information security also aims to ensure that *access to information and the network* is under authorized control and the network is capable of self-defense and countermeasures; the latter is accomplished by specific mechanisms that monitor, detect, react and respond to attacks, vulnerabilities and deficiencies.

Among the most interesting networks in terms of assurance and security are the wireless and the fiber optic. Copper twisted wires are currently used in telephony and for high speed digital at the subscriber loop (DSL).

- *Wireless network* is not a complete end-to-end network; wireless is the access part of the network, which is connected to a traditional network (Internet for data, synchronous for voice and video) and with a traditional medium. With this clarification, there are several wireless networks, each with different security classifications and idiosyncrasies because the transmitted electromagnetic waves reach both friendly and foe antennas. Thus, eavesdropping is a bonus to an intruder as reception is undetectable and security depends on secret keys, terminal ID authentication, and protocols.
  - *The wireless local area network (WLAN)* connects many hundreds of end terminals to a control module, which is connected to a data network server and from there to the data network (Figure 1.7). In the general application of this network, wireless access is point-to-point in a star topology, the wireless link has limited distance (up to 300 m) and the geographic area is limited as well (such as offices on the floor of a building or open space in a campus environment). Wireless LANs are described in the IEEE 802.11 standards [17–26].
  - *The wireless ad-hoc LAN* is a peer-to-peer network. That is, there is no centralized control module, the network topology is meshed, and all end-users maintain connectivity maps continuously, if end-users are mobile, or periodically, if movement is slow or if the network is scalable; wireless sensor networks are among these, as well as mobile cluster end-terminals (for data and voice) that are on a continuous move (Figure 1.8) [18].
  - *The mobile cellular network* is a wireless technology that also provides wireless access to mobile end users. Because the geographic area is partitioned in hexagonal cells, each having its own antenna, the base station (BS), and each having different frequency allocation (known as frequency

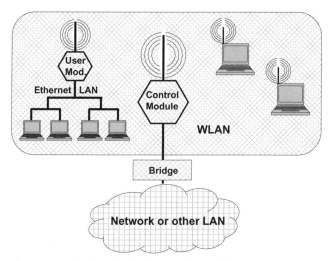

Figure 1.7.   End terminals with a wireless interface access a WLAN via a control module.
Similarly, other LANs (Ethernet) with wireless interface connect with the WLAN. The control module
is connected with the network or another LAN via a bridge that support a high data rate interface
(1GbE or OC-48).

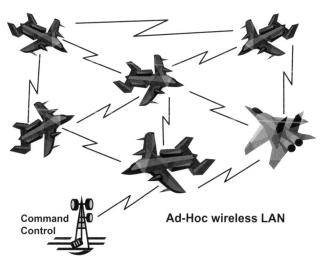

Figure 1.8.   A fully connected (mesh) fast moving and fast reconfiguring ad-hoc wireless network.
Nodes may be added and deleted while the mesh topology is reconfigured. Such applications require
fast and complex protocols with robust security.

reuse) to reduce cross-talk between adjacent cells. Users may move within
a cell and through cells, in which case connectivity changes from base
station to base station according to a protocol known as *handover*. A cluster
of cells communicates with a centralized cell, known as the mobile
telephone switching office (MTSO). The latter communicates with other

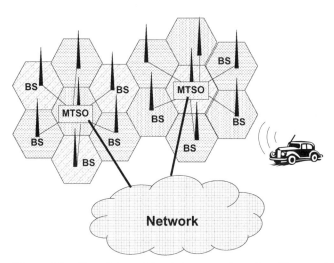

Figure 1.9. The cellular mobile wireless network consists of hexagonal cells in a symmetric cluster; the central cell (MTSO) provides connectivity with the network and with the other surrounding cells (BS) in a star topology.

MTSOs and with the public switched digital network (PSDN) (Figure 1.9) [19].

- ***The fiber optical networks*** currently transport up to 40 Gbps per optical channel, or Tbps per fiber using dense wavelength division multiplexing (DWDM) technology [27]. Although optical technology is more complex than its predecessors, it attracts bad actors because of the huge amount of information that transports in a single fiber. Bad actors with the proper know-how and sophisticated tools may attack the medium (Figure 1.10) and harvest huge amounts of information, mimic the source, alter information or disable the proper operation of the network. Thus, in order to eliminate this risk and assure data security and privacy, highly complex, difficult and sophisticated algorithms are necessary for the generation of the cipher keys and for the key distribution. Moreover, the network itself should be intelligent [28], sophisticated to identify and authenticate channel ID [29], able to detect malicious attackers and outsmart them by adopting sophisticated countermeasure strategies [30] and also vulnerability-free or vulnerability-hardened [31].

1.3.1 ISO/OSI Reference Model and Security

There are many functions in a computer-based node in a data network. These functions have been grouped into abstractions and a hierarchical network reference model of abstraction layers has been defined known as Open System Interconnection (ISO). That is, as information starts from the highest layer and progresses toward the lower layer, specific overhead is added by each layer. Based on this model, a

(photo taken with permission)

Figure 1.10. Photo showing a fiber under tap. The protective PVC tube is cut open, the protective plastic wrap of the fiber cable has been carefully sliced open and a particular fiber has been pulled and tapped. From the tap, another fiber is connected to copying and cryptanalysis equipment. (Photo taken with permission).

Layer		Description
Application	7	•Overall management of a transaction, including segmentation and reassembly
Presentation	6	•Data transformation, formatting and syntax
Session	5	•Negotiated connection parameter •Dialogue control and interrupts •Synchronization mechanisms
Transport	4	•Connection management •Data transfer •Flow control
Network	3	•Move data through the network •Switching and routing functions •Error recovery on network layer
Link	2	•Reliable interchange of data •Error recovery on the link layer •Recovery from abnormal conditions
Physical	1	•Compliance with physical interface specifications •Electrical/optical/wireless, mechanical, functional, procedural

Figure 1.11. The ISO/OSI seven layers model.

node communicates with another peer node of the network and standard protocol (known as peer-to-peer communication) only via its corresponding layer.

All data protocols do not define the same ISO reference model. The International Standards Organization Open Systems Interconnection (ISO/OSI) reference model defines seven layers (Figure 1.11) [32, 33]. The lowest layer (layer 1) is the physical (PHY) and the highest layer (layer 7) is the information. More specifically, the seven layers are defined and contain functions (also known as abstractions). According to the ISO/IEC 7498-1 [5] standard, each protocol layer is composed of three functional planes, the user or bearer, the signalling and control, and the management plane.

The definition of each OSI layer, from the highest to lowest layer, is:

- *Layer 7:* The *Application Layer* deals with communication issues of applications and provides the interface with the user. Examples include the Hyper-Text Transfer Protocol (HTTP), the File Transfer Protocol (FTP), the Session Initiation Protocol (SIP), the Simple Mail Transfer Protocol (SMTP) and Telnet.

- *Layer 6:* The *Presentation Layer* receives data from the Application Layer, translates data, and performs data compression/decompression and data encryption/decryption according to standards. Examples include ASCII, ZIP, JPEG, TIFF, RTP, and the MIDI format.

- *Layer 5:* The *Session Layer* initiates the contact between two computers and sets up the communication lines. It formats the data for transfer and it maintains the end-to-end connection. Examples are the Remote Procedure Call (RPC) and the Secure Sockets Layer (SSL) protocols.

- *Layer 4:* The *Transport Layer* defines how to address the physical locations of the network and to establish connections between host computers, and it handles network messaging. It maintains the session end-to-end integrity and it provides mechanisms to support session establishment for the upper layers. Examples of protocols at this layer are the Transport Control Protocol (TCP), the User Datagram Protocol (UDP), and the Stream Control Transmission Protocol (SCTP).

- *Layer 3:* The *Network Layer* is responsible for routing and relaying data through the network host computers with integrity. It sends *packets* from a source to a destination host computer and it manages bit errors (detection and correction), message routing and traffic control. The Internet Protocol (IP) is performed at this layer.

- *Layer 2:* The *Data Link Layer* defines conditions so that a host can access the network. It ensures message delivery to the proper device over the physical link and it translates, encodes and scrambles the packet bits for the lowest physical layer. Examples at this layer are Ethernet and Token Ring protocol requirements such as 4B/5B or 8B/10B encoding, data scrambling, RZ, NRZI, Manchester, etc.

- *Layer 1:* The *Physical Layer* defines the physical connection between a host computer interface and the network. Its main function is to convert the logical bits received from the data link layer into a physical signal that meets standard transmission specifications, such as min–max voltages or optical impulse power, modulation method (FSK, PSK, etc.) frequency spectrum, bit rate, medium impedance, and so on and for various medium interfaces (wireless, wired, infrared or optical fiber). Hardware drivers at this layer responsible for communication interfaces are network interface cards (NIC).

In addition to grouping functions in layers and defining responsibilities of each layer, ISO [34] and International Telecommunications Union (ITU) standards [35] describe security objectives, services and related mechanisms for each plane and layer. These objectives are accomplished according to a set of criteria, known as *security policies*, which define the provision of security services provided by a layer. Security services are implemented by *security mechanisms*. According to standards, the basic security services are (see also section 1.2.1.3) Authorization, Access Control, Data Confidentiality, Data Integrity, and Non-Repudiation. Similarly, the defined security mechanisms are Encipherment/Decipherment, Digital signatures, Access control, Data integrity, Authentication, Traffic padding, Routing control, Notarization. Security extensions to the security architecture recommended in [35] are provided in [36–53].

1.4 SECURITY THREATENING ATTACKS AND ACTIONS

A communication system that transmits sensitive and proprietary information should be architected and designed to provide the expected security services with secure mechanisms, as described in the previous section. Security services and mechanisms, however, should also vary according to type of network, whether synchronous (circuit switched) or asynchronous (connectionless packet switched), wireless, wired or fiber optic, because the amount of information transported per second as well as node and medium accessibility varies.

In connectionless networks such as the Internet, the first defense is to protect the boundary between the network and the LAN using a firewall device [54, 55]; a firewall runs a special software program that controls, as a single pipe, and examines all the data flow between two hosts or between the data network and the corporate LAN (Figure 1.12). Thus, firewalls can provide a protective mechanism if all configuration parameters have been set correctly and if the running software is the most recent; this occasionally becomes a weak point to new smart attacks that are not recognized by the firewall and thus pass through.

Because of this, security threats may include attacks to the physical medium (tapping wired or fiber optic synchronous networks), or to the node layers (information and/or control of computer-based host) as already described.

Attacking the medium or the information/control layer may be done to enable eavesdropping and copying sensitive data that may be of value to a bad actor, even if data is ciphered (see following section). In such case, the bad actor, by trying different possible keys, may be able to *break* the cipher text. This is known as *brute force attack*. Similarly, a bad actor may capture the key and the cipher text and figure out how to decipher it or how the cryptographic system works. Because attacks of any form have been on the increase, modern networks that are security-minded should include certain additional functions:

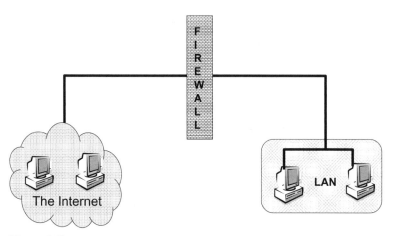

Figure 1.12. A firewall controls the flow of all data traffic from the network to a LAN; if properly configured and running the most recent security software, it is expected to act as a barrier to malevolent attacks and it protects the LAN.

- Detect that an attack by a bad actor took place.
- Be able to differentiate between attacks and malfunctions or faults.
- Locate where the attack took place.
- Detect and verify that the cipher text is not altered or the key is not compromised. If it is suspected that the key has been *compromised*, the key should be *revoked*. In this case, the key establishment may restart and the key may be *updated*.
- Activate a self-defensive countermeasure strategy.
- Activate a counterattack to the bad actor.

Does this sound like a war scenario? It is a war scenario; it is called *cyber-war*.

As a consequence, end-to-end overall security needs to be designed as multi-layer framework that delivers reliability, resiliency, remote access, management, availability and bandwidth and mitigates multi-vendor protocols and interfaces. At the same time, it lowers risk, encompasses services, anticipates and resolves security issues of privacy, intrusion, authentication, policy, and proactive prevention.

1.4.1 Information Security Attacks

Information attacks are alarmingly on an exponential increase. Attacks come in different forms. Among the known viruses are email viruses, Trojan horses, worms, phishing, web pharming, and so on. All have a malicious intention: destroy files, alter files, and harvest and steal sensitive personal data. Many of them may not seem harmful, although they flood the network causing congestion, slow down execution and traffic, cause denial of service, and perhaps can cause routers and the network to shut down. This section examines some types of attacks.

1.4.1.1 *Virus*

A virus is short length software that is attached (piggyback) to a real program, a picture, etc. Each time the program is executed, the virus runs in the background. Depending on the type of virus, when it runs it reproduces copies of itself or clones itself without permission and be attached to other programs while it destroys files, alters the contents of files, or harvests data among specific files. The original virus may modify the reproduced copies, or the copies may modify themselves, known as a metamorphic virus. In such cases, although the initial virus may be detected and destroyed, the modifiefd copies may not be detected.

Such viruses were initially spread from computer to computer via floppy disks. Floppy disks are obsolete and read-only CD-ROMs are used to load programs; CD-ROMs are safer than magnetic media, which can be rewritten and thus modified.

Currently, the most popular method of spreading a virus is through the Internet via email. An email-virus is propagated as an attachment to a message. When activated by clicking, it replicates itself, harvests email addresses from the victim's address book, and automatically sends itself to them, causing damage to all who activate the virus. Thus, a virus may be propagated at an alarming rate and cause havoc. Some email viruses enter a hibernation state and they are activated at a specific time or date, or by a particular triggering event. Other destructive viruses target the boot sector of a rewriteable hard disk. The boot sector contains the first part of the operating system software that controls the computer loading the remaining of the operating system. Modern computers contain the boot program in read-only memory, which cannot be altered by viruses, or the more advanced operating system does not permit other programs to alter the boot sector.

Some viruses are benign; they do not mean to cause damage but to replicate and spontaneously present some text, video, image, or sound. These viruses waste computer resources and execution power, and they slow down the execution of important programs.

Many viruses were written in the scripting languages for Microsoft programs such as Word (.doc), Excel (.xls) or as an executable program (.exe), thus infecting documents and spreadsheets in Microsoft Office or Mac OS based computers. Cross-site scripting viruses use vulnerabilities to propagate, infecting notable sites.

Viruses are also classified as resident and non-resident.

- *Resident viruses* load themselves into memory on execution and transfer control to the host program. The virus stays active in the background and infects new hosts when the infected files are transferred.

- *Non-resident viruses* search for a new target host to infect and they transfer control to the application program they infected.

Anti-virus programs perform a self-integrity check and if they are infected they detect the virus. However, viruses have been programmed to recognize and avoid anti-virus programs. In response to this, anti-virus programs create small bait files that entice viruses to infect them, thus detecting the virus, isolating it, analyzing it, and then destroying it. Thus, bait files are also used as virus forensics tools that examine virus behavior, particularly of polymorphic viruses (or mutating viruses), and to develop improved anti-virus programs.

As anti-virus programs become more sophisticated, so, too, do the viruses. Although they may have infected a file, when the anti-virus program examines the file, the virus sends to the anti-virus program an uninfected portion of the file to trick it. These are known as *stealth viruses*. To counter stealth viruses, anti-viruses employ other methods, and thus the race who tricks whom is on.

Anti-viruses typically scan files to detect specific byte patterns known as virus signatures. When such pattern is detected, the file is placed in quarantine and the user may delete or clean the infected file. If the virus is encrypted, detecting the virus signature becomes more difficult. In this case, the virus is encrypted with a different key for each infected file and only the decrypting portion remains un-encrypted and the same; thus, the anti-virus looks for decrypting modules within files.

1.4.1.2 Trojan Horses

A *Trojan horse* is a computer program that pretends to be one thing (game, movie or sound clip) and in effect it is an executable virus or worm that does harm. The term is borrowed from the historical wooden Trojan horse that was presented by the Greeks to the Trojans at the famous Homeric battle of Troy. According to Homer, many Greek soldiers were hidden in the belly of the wooden horse, after the Trojans brought the horse inside the wall of the citadel, during the night the Greeks opened a trap door, descended from the horse and opened the wall gates. The fall of the city and the disaster that followed it brings to mind what the computer Trojan horse can accomplish. Trojan horses are also known as *security loopholes*.

Typically, Trojan horses exploit social engineering and propagate via email hiding behind legitimate animations, screensavers, or some other fun files (a typical Trojan horse program has the .**src** extension). When the user clicks on the fun file to watch it, the malicious program is installed and it runs its destructive course. A Trojan horse may affect the registry by adding entries so that the virus runs each time the computer boots, or by adding services to the computer, opening ports and allowing hackers to remotely access the user's computer (called a RAT, remote administration tool), disable security software, destroy files, cause denial of service, affect the downloader, send email, upload or download files, harvest email addresses, stealing passwords, spreading viruses, phishing for bank accounts, turn off the computer, and more.

Trojan horses are not known to replicate themselves. Instead, they reside on the host computer and propagate from computer to computer when an email with the virus file is transmitted to another host computer.

There is no single method to delete Trojan horses. The simplest method is to clear all temporary Internet files on the computer, or edit the registry. It may even be necessary to reset the computer back to its factory defaults. The latest version of reputable anti-virus software is strongly recommended for all virus types including Trojan horses.

1.4.1.3 Worms

A worm is a very short length software (few hundreds of lines) that exploits software security vulnerabilities (known as *security holes*) in a computer; if it finds a hole, it replicates itself. Copies of the worm propagate across the computer network and when a worm finds a computer with similar holes, starts replicating there as well, and so on. Effectively, this results in an avalanche effect of worm propagation across the computer network that causes considerable harm. Thus, worms waste computer execution time and considerable network bandwidth that could clog the network, effectively causing the network to slow or shut down.

For example, the worm dubbed *Code Red* (July 2001), on July 19, 2001 replicated itself more than 250,000 times in approximately nine hours. This worm infected computers causing web pages to be replaced by a page displaying the message "Hacked by Chinese." It was designed to replicate itself 20 days of each month, and it attacked the White House; it made 100 simultaneous attempts to port 80 of the address www.whitehouse.gov (198.137.240.91). This address has changed since then and a security patch was issued.

In 2003, the *Slammer worm* exploited a buffer overflow vulnerability. It generated random Internet addresses (it infected hosts over the UDP) and it sent out copies of the worm to those addresses. Because the worm was small (376 bytes), it fit in a single small packet and it was routed faster than longer packets (some routing algorithms push short packets through faster than longer ones). The worm clog that was generated caused some routers to shut or to slow down, creating severe congestion. Neighboring routers, which were trying to reroute traffic and avoid the congested routers, started sending messages to other computers in the network to update their routing tables, causing more traffic over the network. The avalanche of worms and table updates flooded the network, eventually causing data traffic to significantly slow down or to be dropped.

In 2007 the *Storm* worm was launched. Storm used social engineering and tried to entice users to load free music from well-known artists. The worm hid in the music files. The worm in the infected computer (called *zombie* or *bot*) opened a back door, adding the Microsoft Window-based OS computer to a peer-to-peer group of computer networks called a *botnet* without the consent of the computer user; the term botnet derives from ro*bot-net*work. Each group of computers in a botnet may consist of few hundreds to several thousands, and therefore it's estimated that many millions of computers may have been infected with the Storm worm. The originator of a botnet, known as a *botnet herder*, may remotely control the botnet.

Although the initial motivation of a botnet was not malicious, and therefore botnets comply with RFC 1459 [56], botnets were exploited and evolved to accommodate malicious intentions and compromise computers. Thus, *bots* on the botnet deliver spam or adware (estimated in the millions or billions), attack websites, and harvest computer data and resources, thus wasting bandwidth and causing denial-of-service [57]. Harvesting computer resources is also known as "scriumping."

1.4.1.4 Phishing and Pharming

Phishing is an illegal attempt via email or instant messaging to entrap someone to divulge personally sensitive information using social engineering methods. Such an attempt may be impersonating a trustworthy bank and asking for bank accounts; masquerading as a credit card service provider and asking to verify the credit card number and pin; or masquerading as a trustworthy entity (police, technical, education, training, legislative, web service, and so on) and asking for specific information (username and password, health status, personal data, social security number, etc). The term *phishing* derives from fishing, whereby the impersonator lures the user to fall victim.

Here is a real example of phishing that the author received on November 7, 2007 (at 10:47 EDT):

Security Alerts
RBC Royal Bank [ibanking@app.rbc.com]

To: Kartalopoulos, Stamatios V.

Cc: _____

Dear Customer,
Our Technical Service department has recently updated our online banking services, and due to this upgrade we sincerely call your attention to follow below link and reconfirm your online account details. Failure to confirm the online banking details will suspend you from accessing your account online.

https://www1.royalbank.com/cgi-bin/rbaccess/rbunxcgi

We use the latest security measures to ensure that your online banking experience is safe and secure. The administration asks you to accept our apologies for the inconvenience caused and expresses gratitude forcooperation.

Thanks for banking with us.

Royal Bank of Canada Security Advisor
RBC Financial Group
_____

This is an automatic message. Please do not reply.

_____

Phishing is a very serious offense and in some ways it can be more harmful than a worm or virus. If the victim believes that the phishing email comes from a bank, and thus gives away a bank account and other sensitive personal data the phishing culprit may steal the user's identity and wealth. Thus, one anti-phishing defense is for users to be alert and to not give personal data over email. In general, legitimate and trustworthy entities do not solicit personal information via email or instant messaging. Banks and government agencies do not, as a rule, ask via email for accounts and social security numbers.

The Anti-phishing Act of 2005 [58] is a bill that was referred to the U.S. Senate and also to the Subcommittee on Crime, Terrorism, and Homeland Security of the U.S. House of Representatives to combat phishing and pharming, and to define penalties for individuals who commit such crimes and identity theft by falsifying corporate and other entity websites or emails. Similarly, in the United Kingdom, a general offense of fraud was introduced with the Fraud Act 2006 [59]. Despite these two bills and many others, the computer user should also be prudent and make a serious effort to combat phishing and pharming by using the latest versions of spam filters.

Phishing emails use social engineering tactics that aim to upset and panic the recipient of the email, particularly the elderly. For example, this bogus email has specific key-words that can upset and disturb the recipient:

From: 1st United Services Credit Union [mailto:services@1stuscu.org]
Sent: Mon 11/12/2007 10:12 AM
Subject: Important Member Service Information

Dear Member,

This is your official notification from 1st United Services Credit Union that the service(s) listed below will be deactivated and deleted if not renewed immediately.
Previous notifications have been sent to the Billing Contact assigned to this account.
As the Primary Contact, you must renew the service(s) listed below or it will be deactivated and deleted.
Renew Now your **1st United Services Credit Union and ON-LINE BANKING.**
SERVICE: **1st United Services Credit Union ON-LINE BANKING.**
EXPIRATION: **November, 16 2007**
Thank you for using 1st United Services Credit Union.
We appreciate your business and the opportunity to serve you.
1st United Services Credit Union ON-LINE BANKING

Copyright © 2007 1st United Services Credit Union. All rights are reserved.

Examination of this message reveals the following tricks of appearance and intimidation designed to induce panic:

- First, it attempts to look like a legitimate notice with a fake copyright at the end of the email. However, there is no "to:" and it is addressed to an impersonal "member" and not to a specific name.
- It is addressed to the "Primary Contact" and states that the recipient "must renew the service(s) listed below," but there is no list of services. Instead, it entices the recipient you to click on the "Renew now" to look for services, etc., and get trapped.
- It uses the strong words and it gives only 5 days margin to renew or "the account will be deactivated and deleted immediately"

Pharming (a derivative of farming) also attempts to illegally collect personal sensitive information by *domain spoofing*. That is, instead of using bogus email requests, pharming "poisons" a domain node server (DNS) by infusing false information in it and redirecting a website's traffic to another bogus website which harvests personal data from the user's computer; as far as the browser is concerned, the user is connected with the correct website.

DNS servers are responsible for resolving Internet names into their real addresses. For example, the user types a name such as Amazon.com, AOL.com, Google.com, etc, and the DNS server translates it to a numeric address.

A key difference between phishing and pharming is that the first targets one user at a time via email whereas the second targets large groups of users through domain spoofing; spoofing uses someone else's IP address in TCP/IP packets (source or destination address). With spoofing, the attacker pretends that is the source or the destination and may cause redirection of routing to the hacker's host, or may flood the receive queues causing a large group of users to respond to the victim. Pharming has become a major concern to e-commerce.

1.4.1.5 Protecting the Computer from Viruses

The best defense against viruses, Trojan horses, worms and other known attacks is to install the latest version of reputable anti-virus and anti-spam security software. In addition, the user may also take few simple protection steps, such as:

- Saving files periodically on an external had disk. If the computer is infected, the user can always use back-ups on a clean computer.

- Not opening emails that do not seem legitimate, either because the subject does not read right, or the sender does is not familiar, or it promises some free fun adventure. Remember, no one gives anything for nothing. Remember that bad actors are unscrupulous and very sophisticated.

- If an email has an attachment of an executable file (.exe, .com, .vbs, .src), never open it. In general, do not open attachments from sources that seem suspicious. Even a picture (.gif) or video clip may contain a virus, a worm or a Trojan horse. Ask, what is more valuable? The attachment or the files in the computer?

- Install software programs from CD-ROMs only. Reputable software companies have enough security safeguards and the software they sell comes on a one time writeable disk, the CD-ROM, which subsequently cannot be infected by viruses. Thus, avoid programs from online, unknown sources.

- The Macro Virus Protection of the computer should be enabled and in general, users should avoid running macros in a document.

1.4.2 VPN Networks

1.4.2.1 Tunneling

Tunneling is a term that encompasses encapsulation, routing, and decapsulation. Encapsulation entails forming a new packet by adding header overhead to the original packet. The overhead of the new packet may have a new address and routing information that is used to route the new packet through a logical data path. To the original source, the tunnel appears as a point-to-point transparent connection across the network and is unaware of routers, switches, security gateways and proxy servers on the tunnel path. During the encapsulation process, the original packet can be encrypted for confidentiality purposes. When tunneling is combined with confidentiality, it can be used to provide a virtual private network (VPN). Encapsulated packets travel through the network tunnel. When the encapsulated packet reaches its destination at the end of the tunnel, the encapsulation overhead is removed and the destination information on the original packet is used to route the packet to its final destination. That is, routing becomes a two-step process.

1.4.2.2 VPN Security

A Virtual Private Network (VPN) is a private network, which by configuring a public data network like the Internet emulates a point-to-point private secure link between two sites (site-to-site) or between a remote client computer and a corporate server; this is also known as tunneling. Thus, two corporations at different sites and clients who travel or who work from home are able to obtain remote access connectivity between their VPN computer (VPN client) and an organization server (corporate VPN server) using provisioned links of the public data network infrastructure (such as the Internet). Based on this, there are two types of VPN connections:

- *Site-to-site* by which a router establishes VPN connections through the public data network between two private networks.

- *Remote access* by which a remote access client makes a remote access VPN connection with a private network.

VPN configuration for the initial connection is common to both types and this includes tunneling protocols, authentication methods, access network and address assignment.

Typically, servers provide several VPN protocols for remote client access connections and for site-to-site connections. Among them are the Point-to-Point Tunneling Protocol (PPTP), the Layer Two Tunneling Protocol over Internet security (L2TP/IPsec), and the IPsec tunnel mode.

- ***The Point-to-Point Tunneling Protocol (PPTP)*** supports on-demand, multiprotocol, and it enables the secure transfer of data from a remote client to a private corporate server by creating a VPN link across TCP/IP-based data networks. PPTP connections require only user-level authentication through a point-to-point authentication protocol.

- *The Layer Two Tunneling Protocol (L2TP)* provides encapsulation for Point-to-Point Protocol (PPP) frames across packet-oriented (IP) networks, and it allows IP traffic to be encrypted. L2TP may use Internet Protocol security (IPsec) encryption to protect the data stream over the link between the VPN client and the VPN server. In this case, it is termed L2TP over IPsec (L2TP/IPsec). L2TP/IPsec connections require the same user-level authentication with PPTP and, in addition, computer-level authentication using computer certificates.

- *The IPsec tunnel mode* allows IP datagrams to be encrypted, encapsulated by adding IP header, and sent across a corporate IP network or a public IP network. In this mode, IPsec provides encapsulation for IP traffic only. IPsec tunnel mode provides interoperability with routers, gateways and end systems that do not support L2TO/IPsec or PPTP VPN tunneling (Figure 1.13).

In some applications, the VPN server is integrated into the firewall functionality, from which site-to-site VPN connections as well as VPN client access to the corporate network can be managed.

In VPN point-to-point links, data is encapsulated with a header that includes routing information through the public network to its destination address. For confidentiality, data is also encrypted.

A call from a remote client (or gateway) to establish VPN connectivity entails several steps:

- A remote access client (or remote site gateway) dials the server.

- The server sends a challenge to the client (according to an authentication protocol, such as CHAP or MS-CHAP).

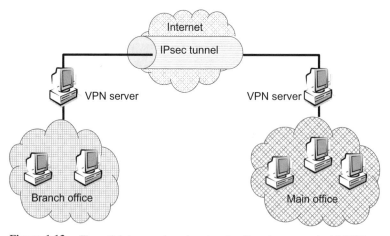

Figure 1.13. The path between a branch and main offices is connected with VPN servers and IPsec tunnel mode through the Internet network.

Table 1.3. Security comparison of certain data protocols

Protocol	Usage	Security
PPTP	Connecting to VPN server	Moderate
L2TP/IPsec	Connecting to VPN server	High
IPsec Tunnel mode	Third-party VPN server	High

- The client (or gateway) responds to the challenge and sends an encrypted response to the server that consists of a user name, a domain name, and a password; these credentials may differ according to the authentication protocol in use.
- The server checks the response against a valid user account database.
- If (according to the database) the client response is authenticated, the server uses the dial-in properties of the user account and the remote access policies to authorize the connection.
- If the client does not respond with acceptable credentials, the access attempt fails.

These steps repeat each time the client attempts to access a network resource.

While a remote VPN client has requested connectivity and until the credentials have been verified, the client may have been placed under quarantine control. This determines that the client's computer configuration is in accordance with the organization's specific quarantine restrictions (type of anti-virus software) and VPN policy. If the client fails to meet configuration requirements and if they cannot be corrected, the client fails to connect after some predetermined interval. Quarantine control may be optional.

A comparison of three VPN protocols is provided in Table 1.3.

1.4.2.3 IPsec

The Internet Protocol Security or IPsec is a suite of security protocols designed to secure IP traffic between computers. As already discussed, a secure channel starts with two peer computers, one at each end of the channel, which define a secret session key known as authenticated key exchange. Then, the peers encrypt the information over the channel via MAC protocols (such as HMAC).

IPsec is designed by the Internet Engineering Task Force (IETF) as the security architecture for the Internet Protocol (IP) to provide interoperable cryptographically-based security for IPv4 and IPv6.

IPsec defines (at the IP layer) IP packet formats to provide end-to-end (client-to-client, client-to-server, server-to-server) security services such as access control, connectionless integrity, data origin or data source authentication (filtering incoming

IP addresses to eliminate spoofing), integrity, anti-replay (detecting and rejecting partial sequence integrity), confidentiality via encryption, and limited traffic flow confidentiality.

Based on this, the MACs of the two (end-to-end) peers authenticate each other and with IPsec the receiving MAC is assured that the source IP address was initiated at the sourcing MAC, although there is no warranty that the message is that of the sourcing terminal and that a third party has not mimicked the source. That is, IPsec protects communication between computers but it is not involved in the user authentication or authorization between VPN and user-terminal.

The IPsec suite of protocols consists of several protocols, two of which are for channel authentication, and they are cryptographic algorithm-independent. That is, they are modular, permitting selection of different cryptographic algorithms and they can run either independently or one on top of the other to provide IPv4 and IPv6 security:

- *The Authentication Header (AH)* protocol provides data origin authentication and integrity protection but it does not encrypt the channel.

- *The Encapsulating Security Payload (ESP)* protocol provides the same services AH does, and it provides confidentiality and limited traffic flow confidentiality for concealing packet length and facilitating generation and discard of dummy packets; ESP requires more bandwidth.

Both AH and ESP protocols support two distinct modes of operation: transport and tunnel. To get the channel ready for communication, a key exchange authentication is performed as described by two protocols, the Internet Security Association Key Management Protocol (ISAKMP) and the Internet Key Exchange (IKE). IKE also helps to establish a simplex (unidirectional) Security Association (SA) between two computers, which is defined by a peer's IP address, AH or ESP, and an index to a set of parameters such as the encryption and hash algorithms.

IPsec creates a boundary between unprotected and protected interfaces and traffic traversing the boundary is subject to access controls. As such, IPsec provides secure gateway-to-gateway connections across private wide area networks (WAN), internet-based connections using Layer Two Tunneling Protocol over IPsec tunnels (L2TP/IPsec), or pure IPsec tunnel mode; the latter is not designed for VPN remote access. Additionally, IETF defines on-demand security negotiation and automatic key management service.

The Point-to-Point Tunneling Protocol (PPTP) and L2TP/IPsec are among the tunneling protocols that establish a tunnel between two VPN servers (VPN site-to-site). Instructions how to set up IPSec remote networks are provided by the manufacturer. Typical steps are:

- Select IPsec as the tunneling protocol of the remote site network.
- Configure the VPN end-servers in IPsec tunnel mode and the advanced IPsec settings based on instruction and access policy. Caution: some VPN servers

will erase the IPSec configuration when a restart is invoked and all traffic from clients may be forwarded to the Internet unencrypted.

In VPN site-to-site with IPsec if L2TP over IPsec is used, the IPsec tunnel (shown in Figure 1.12) is replaced by L2TP/IPsec and the VPN servers are configured accordingly, following manufacturer's instructions.

Althought IPsec consists of a large suite of protocols [60–78] to provide the required level of detail and protocol inter-relationship, its understanding and implementation is not as easy. For example: different RFCs describe the AH and ESP protocols, cryptographic algorithms for integrity and encryption for use with AH, ESP and different IKE versions.

1.4.3 Network Security Attacks

1.4.3.1 Network and Service Availability

Network availability is a primary requirement of a communication system. Availability means that the system should be ready, within a specified and tight margin, to provide the expected services requested by customers under any state of the network, whether normal, congested, or under attack condition. That is, the network must be available to provide the agreed-upon services whenever the user needs it.

However, availability of network and of services depend on network technology, architecture, design, medium, protocols, topology and survivability tactics. Because networks are dissimilar, security of each network is addressed with different mechanisms. For example, wireless ad hoc networks and asynchronous connection-less networks do not have centralized control in contrast to mobile cellular and to synchronous circuit switched networks, and thus availability of each network is not expected to be the same. Similarly, a residential access network (wired or fiber) is not expected to have the same availability with a backbone mesh fiber optic network. Thus, the vulnerabilities of each network are not the same.

Network, protocol and service vulnerability is important to security of both information and network. Bad actors exploit vulnerabilities to harvest information, attack the network to cause denial of service, or bring down the network. Thus, a unified security approach is sought. In fact, because bad actors become more aggressive in their attacks, the modern network should be more intelligent, self-defensive, self-protective and in many cases counter-attacking in order to protect itself and be survivable and available for user service.

1.4.3.2 Network Attacks

Network attack encompasses any malicious actions that aim to intentionally interrupt, disturb, slow down, stop, or cause the network to malfunction or operate at degraded performance. There are two classifications of network attacks: passive and active.

- *Passive attacks* are considered those that do not directly affect the medium or the network. For example, in wireless a suitably-equipped eavesdropper is "listening" to conversations harvesting information, authentication, ID codes, pin numbers, and encryption keys, as well as networking information such as OA&M (operation, administration and management) that can be used at a later time to launch an assault.

- *Active attacks* are considered those that directly impact the transmission characteristics of the network. For example, in wireless access networks the attacker may broadcast a multi-frequency signal to jam or corrupt communication channels. In wired or optical networks, that attacker may intercept traffic by tapping the medium or the switch and eavesdrop, mimic the source of data, alter data, reverse engineer the cryptographic system (that is crypto-analyze cipher-texts), derail routing, affect the control, operations and management of nodes, affect billing data, and also cause nodes to not cooperate with the protocols and expected behavior in the network.

Both types of attacks may be attempted by a disgruntled employee or an unauthorized person inside communications buildings (central office, communications vault or closet), or inside an outdoor cabinet or hut that houses communications equipment. Attacks may also be outside the premises; such as attacking cable links (tapping aerial cables or cables in underground pipes and pathways and in manholes).

Among such attacks, the most typical are:

- *Eavesdropping* attempts to "listen" to data transported over a medium (wireless, wired, or optical) and to copy data, keys, protocols, and any sensitive information that can be useful to a bad actor.

- *Interception* implies capturing a message and either destroying it or altering and retransmitting it by mimicking the source.

- *Impersonation* is also source-mimicking, although it does not imply that a message must have been intercepted. Impersonation may source false original data to a destination in an attempt to mislead the receiver, or to engage the receiver in a cryptographic protocol process which can facilitate the bad actor's cryptanalysis work.

- *Routing* are attacks that aim the routing tables and protocols of nodes in the network, including provisioned tunnels. Such attacks may aim to derail specific channels to other destinations, to cause (undesirable) broadcasting of data, and to cause severe congestion and denial of service over specific paths or node-clusters of the networks.

In conclusion, network attacks are on several layers, particularly on the information, control and physical layer, and a network cannot be considered secure if only one or two of the three are considered secure and not all three.

The best cryptography does NOT warranty secrecy if nodes are not immune to attacks; and the most immune node does NOT warranty privacy if the network is NOT secure.

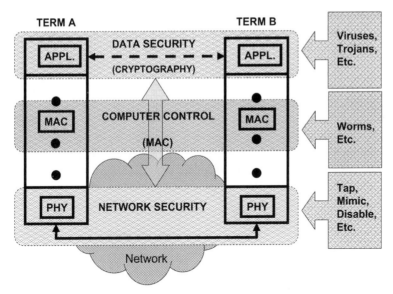

Figure 1.14.   The three most vulnerable layers according to the ISO model.

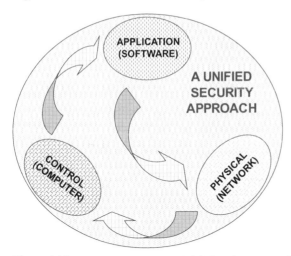

Figure 1.15.   A synergistic approach definitely enforces network security. This synergy may also be used to architect self-defensive as well as counter-attacking networks.

Figure 1.14 addresses the three particular ISO layers that are most vulnerable to attacks—information, control and physical—and Figure 1.15 encapsulates a synergistic approach to security.

1.4.3.3 *Counter-Attacking Intelligent Networks*

Counter-attacking is best understood based on the scenario of a bad actor attacking the medium.

- First, this attack is detected as it occurs. Chapter 7 discusses methods that do exactly this; that is, they detect channel attacks in real-time and with no service interruption by the method.

- As soon as the attack is detected, the nodes at either side of the link initiate a counter-measure strategy, whereby sensitive data is moved to another secure channel and whereas decoy data are being transmitted over the attacked channel.

- In addition, since the attacker operates with computer-based equipment, the decoy messages contain hidden viruses and Trojan horses, which although not destructive to link receivers (by preconditioning the receivers), are destructive to the bad actor's computer.

REFERENCES

1. Herodotus, *Histories*, (many publishers).
2. Homer, The *Illiad* (many Publishers and translations).
3. Aeschylus, *Agamemnon*, Loeb Classical Library.
4. D. KAHN, "Seizing the Enigma: The Race to Break the German U-Boats Codes, 1939–1943" (1991).
5. A. STRIPP, "The Enigma Machine: Its Mechanism and Use", in Hinsley and Stripp (eds.) *Codebreakers: The Inside Story of Bletchley Park*, 1993, pp. 83–88.
6. FIPS 180-1, "Secure Hash Standard", National Institute of Standards and Technology, April 17, 1995.
7. FIPS 186-2, "Digital Signature Standard", National Institute of Standards and Technology, 15 February 2000.
8. FIPS Pub 198, The Keyed-Hash Message Authentication Code (HMAC), March 2002.
9. FIPS Pub 190, *Guideline for the use of advanced authentication technology alternatives*, September 28, 1994.
10. IEEE P1363, "Standard Specifications for Public Key Cryptography", Institute of Electrical and Electronics Engineers, 2000.
11. W. DIFFIE and M. HELLMAN, "New Directions in Cryptography", *IEEE Trans. Info. Theory*, IT-22, 1976, pp. 644–654.
12. FIPS Pub 185, Escrowed Encryption Standard, February 9, 1994.
13. CRS Report to Congress, Order Code RL33123, *Terrorist Capabilities for Cyberattack: Overview and Policy Issues*, January 22, 2007.
14. CRS Report to Congress, Order Code RL32114, *Computer Attack and Cyberterrorism: Vulnerabilities and Policy Issues for Congress*, April 1, 2005.
15. CRS Report to Congress, Order Code RL32331, *The Economic Impact of Cyber-Attacks*, April 1, 2004.
16. CRS Report to Congress, Order Code RL31542, *Homeland Security-Reducing the Vulnerabilitiy of Public and Private Information Infrastructures to Terrorism: An Overview*, December 12, 2002.
17. IEEE 802.11, 1999 Edition, Part 11: Wireless LAN Medium Access Control (MAC) and Physical Layer (PHY) Specifications.
18. IEEE 802.11a-1999, Part 11: Wireless LAN Medium Access Control (MAC) and Physical Layer (PHY) specifications—Amendment 1: High-speed Physical Layer in the 5 GHz band.
19. IEEE 802.11b-1999, Supplement to 802.11-1999, Wireless LAN MAC and PHY specifications: Higher speed Physical Layer (PHY) extension in the 2.4 GHz band.

20. 802.11b-1999/Cor1-2001, Part 11: Wireless LAN Medium Access Control (MAC) and Physical Layer (PHY) specifications—Amendment 2: Higher-speed Physical Layer (PHY) extension in the 2.4 GHz band—Corrigendum1.

21. IEEE 802.11d-2001, Amendment to IEEE 802.11-1999, Part 11: Wireless LAN Medium Access Control (MAC) and Physical Layer (PHY) Specifications: Specification for Operation in Additional Regulartory Domains.

22. IEEE 802.11e-2005, Part 11: Wireless LAN Medium Access Control (MAC) and Physical Layer (PHY) specifications: Amendment 8: Medium Access Control (MAC) Quality of Service Enhancements.

23. IEEE 802.11g-2003, Part 11: Wireless LAN Medium Access Control (MAC) and Physical Layer (PHY) specifications—Amendment 4: Further Higher-Speed Physical Layer Extension in the 2.4 GHz Band.

24. IEEE 802.11h-2003, IEEE Standard for Information technology—Telecommunications and Information Exchange Between Systems—LAN/MAN Specific Requirements—Part 11: Wireless LAN Medium Access Control (MAC) and Physical Layer (PHY) Specifications: Spectrum and Transmit Power Management Extensions in the 5 GHz band in Europe.

25. IEEE 802.11i-2004, Part 11: Wireless LAN Medium Access Control (MAC) and Physical Layer (PHY) specifications–Amendment 6: Medium Access Control (MAC) Security Enhancements Interpretation.

26. IEEE 802.11j-2004, Part 11: Wireless LAN Medium Access Control (MAC) and Physical Layer (PHY) specifications—Amendment 7: 4.9 GHz–5 GHz Operation in Japan.

27. S.V. KARTALOPOULOS, *DWDM: Networks, Devices and Technology*, Wiley/IEEE Press, 2003.

28. S.V. KARTALOPOULOS, *Next Generation Intelligent Optical Network: From Access to Backbone*, Springer, 2007.

29. S.V. KARTALOPOULOS, "Channel signature authentication for secure optical communications", SPIE Defense & Security, 9/26–29/05, Bruges, Belgium, paper no: 5986-41, on CD-ROM: CDS191.

30. S.V. KARTALOPOULOS, "Optical Network Security: Countermeasures in view of channel attacks", Unclassified Proceedings of Milcom 2006, October 23–25, 2006, Washington, D.C., on CD-ROM, ISBN 1-4244-0618-8, Library of Congress 2006931712, paper no. US-T-G-404.

31. S.V. KARTALOPOULOS, "Identifying vulnerabilities of quantum cryptography in secure optical data transport", Unclassified Proceedings of Milcom 2005, 10/17–20/05, Atlantic City, session: Comm. Security I, invited paper # 678, on CD-ROM, ISBN # 0-7803-9394-5.

32. ISO/OSI 7498-1: 1984, *Open Systems Interconnection—Part 1: Basic Reference Model,*; also ISO/IEC 7498-1, 1994.

33. ITU-T Recommendation X.200, *Open Systems Interconnection—Basic Reference Model: The basic model*, 1994.

34. ISO/IEC 7498-2, *Open Systems Interconnection—Part 2: Security Architecture*, 1989.

35. ITU-T Recommendation X.800, *Security Architecture for Open Systems Interconnection for CCIT Applications*, 1991.

36. ITU-T Recommendation X.802, *Open Systems Interconnection—Lower Layers Security Model*, 1995.

37. ITU-T Recommendation X.802, *Open Systems Interconnection—Upper Layers Security Model*, 1994.

38. ITU-T Recommendation X.810, *Open Systems Interconnection—Security Frameworks for Open Systems: Overview*, 1995.

39. ITU-T Recommendation X.811, *Open Systems Interconnection—Security Frameworks for Open Systems: Authentication Framework*, 1995.

40. ITU-T Recommendation X.812, *Open Systems Interconnection—Security Frameworks for Open Systems: Access Control Framework*, 1995.

41. ITU-T Recommendation X.813, *Open Systems Interconnection—Security Frameworks for Open Systems: Non-repudiation Framework*, 1996.

42. ITU-T Recommendation X.814, *Open Systems Interconnection—Security Frameworks for Open Systems: Confidentiality Framework*, 1995.

43. ITU-T Recommendation X.815, *Open Systems Interconnection—Security Frameworks for Open Systems: Integrity Framework*, 1995.

44. ITU-T Recommendation X.815, *Open Systems Interconnection—Security Frameworks for Open Systems: Security Audit and Alarms Framework*, 1995.

45. ISO/IEC 10181-1: 1996, *Open Systems Interconnection—Part 1: Security Frameworks for Open Systems: Overview*.

46. ISO/IEC 10181-2: 1996, *Open Systems Interconnection—Part 2: Security Frameworks for Open Systems: Authentication Framework*.

47. ISO/IEC 10181-3: 1996, *Open Systems Interconnection—Part 3: Security Frameworks for Open Systems: Access Control Framework*.

48. ISO/IEC 10181-4: 1996, *Open Systems Interconnection—Part 4: Security Frameworks for Open Systems: Non-repudiation Framework*.

49. ISO/IEC 10181-5: 1996, *Open Systems Interconnection—Part 5: Security Frameworks for Open Systems: Confidentiality Framework*.

50. ISO/IEC 10181-6: 1996, *Open Systems Interconnection—Part 6: Security Frameworks for Open Systems: Integrity Framework*.

51. ISO/IEC 10181-7: 1996, Open Systems Interconnection—Part 7: Security Frameworks for Open Systems: Security Audit and Alarms Framework.

52. ISO/IEC 11586-1: 1996, Open Systems Interconnection—Generic upper layers security: Overview, models and notation.

53. ISO/IEC 11586-2: 1996, *Open Systems Interconnection—Generic upper layers security: Exchange Service Element (SESE) service definition*.

54. NIST 800-41, *Guidelines for Firewalls and Firewall Policy*, January, 2002.

55. GAO-04-467, United States General Accounting Office, Report to Congressional Requesters, *INFORMATION SECURITY, Technologies to Secure Federal Systems*, March 2004.

56. RFC 1459, Internet Relay Chat Protocol, May 1993.

57. Sharon Gaudin, "Storm Worm Botnet More Powerful Than Top Supercomputers", Information Week, September 6, 2007. http://www.informationweek.com/news/showArticle.jhtml?articleID= 201804528

58. Anti-Phishing Act of 2005, "*A bill to criminalize Internet scams involving fraudulently obtaining personal information, commonly known as phishing*", March 1, 2005; http://www.govtrack.us/ congress/billtext.xpd?bill=h109-1099

59. Elizabeth II, UK, Fraud Act 2006, Chapter 35, 8th November, 2006.

60. RFC 2401 "Security Architecture for the Internet Protocol".

61. RFC 2402, "IP Authentication Header".

62. RFC 2451, "The ESP CBC-Mode Cipher Algorithms".

63. RFC 2403, "The Use of HMAC-MD5-96 within ESP and AH".

64. RFC 2404, "The Use of HMAC-SHA-1-96 within ESP and AH".

65. RFC 2405, "The ESP DES-CBC Cipher Algorithm With Explicit IV".

66. RFC 2406, "IP Encapsulating Security Payload (ESP)".

67. RFC 2407, "The Internet IP Security Domain of Interpretation for ISAKMP".

68. RFC 2408, "Internet Security Association and Key Management Protocol (ISAKMP)".

69. RFC 2409, "The Internet Key Exchange (IKE)".

70. RFC 2857, "The Use of HMAC-RIPEMD-160-96 within ESP and AH".

71. RFC 3526, More Modular Exponential (MODP) Diffie-Hellman groups for Internet Key Exchange (IKE)".

72. RFC 3554, "On the Use of Stream Control Transmission Protocol (SCTP) with IPsec".

73. RFC 3566, "The AES-XCBC-MAC-96 Algorithm and Its Use With IPsec".

74. RFC 3602, "The AES-CBC Cipher Algorithm and Its Use with IPsec".

75. RFC 3664, "The AES-XCBC-PRF-128 algorithm for IKE".
76. RFC 3686, "Using AES Counter Mode With IPsec ESP".
77. RFC 3706, "A Traffic-Based Method of Detecting Dead IKE Peers".
78. RFC 3715, "IPsec-NAT Compatibility Requirements".

Chapter 2

Mathematical Foundations

2.1 INTRODUCTION

This chapter reviews specific topics that are most commonly used in cryptography. In a serious effort to present these topics for non-mathematicians, (advanced mathematical notation) has been eliminated wherever possible. Among the topics are logarithms, prime numbers, modulus arithmetic, greatest common divisor, groups, rings, fields, and Exclusive-Or.

2.2 LOGARITHMS

Logarithm is a compacted mathematical form that converts calculations with exponents that yield large numbers to smaller numbers performing addition or subtraction. According to this, the expression $b^y = x$ is converted to $y = \log_b x$, where b is the base of the logarithm. For example, $10^3 = 1000$ is written with logarithms as $\log_{10} 1000 = \log_{10} 10^3 = 3$, and $2^8 = 256$ is written as $\log_2 256 = \log_2 2^8 = 8$. The latter example, numbers expressed as powers of 2 and using logarithm base 2, is more popular in binary arithmetic as well as in cryptography. Some of the most commonly used logarithmic properties are:

$$\log(ab) = \log a + \log b$$

$$\log(a/b) = \log a - \log b$$

$$\log(a^b) = b \log a$$

$$\log_b(b) = 1$$

$$\log(1) = 0$$

$\log(0) =$ not defined

$\log(a + b) = \log(kn) = \log k + \log n$, where k and n are two numbers such that $a + b = kn$,

or

$\log(a + b) = \log(k^n) = n\log k$, where k and n are two numbers such that $(a + b) = k^n$; for example, $\log(6 + 3) = \log(3^2) = 2\log 3$.

In binary calculations and for brevity it is common not to state the base, as in $\log 2^8 = 8$. Caution: if the base is assumed to be 2 and is not stated, one should not mix logarithms with different bases. However, a logarithm of one base x can be converted to a logarithm of another base y using the conversion relationship $\log_x a = \log_y a / \log_y x$.

Using the most common base logarithms, base 2 and base 10:

$$\log_2 a = \log_{10} a / \log_{10} 2, \text{ where } \log_{10} 2 = 0.301029995 \text{ and } \log_{10} 10 = 1.$$

For example:

$$\log_2 10 = \log_{10} 10 / \log_{10} 2 = 1/0.301029995 = 3.341928102$$

And also:

$$\log_2 32 = \log_{10} 32 / \log_{10} 2 = \log_{10} 32 / \log_{10} 2 = 1.50514990/0.30102999 = 5.0000000$$

2.3 PRIME NUMBERS

Prime numbers have been the subject of investigation since antiquity. Both ancient and modern mathematicians have been intrigued by them because of their elegance and their difficulty to find and manipulate. Their difficulty is of main importance in cryptography where they have found a widespread applicability; there are still many unsolved problems with prime numbers [1].

The philosopher-mathematicians of the Pythagorean's school (500 BCE to 300 BCE) were interested in the harmonic relationship of numbers, believing that their harmony is linked with cosmic symmetry, balance and relationship. From their studies, they were able to link the relationship of numbers with the harmony of sound (music) and colors, as well as the relationship of planets and constellations. They also understood *perfect* numbers, *amicable* numbers and the idea of *primality*.

- A *perfect number* is one whose proper divisors sum to the number itself. The proper divisors 1, 2 and 3 of number 6 sum up to 6, and the proper divisors 1, 2, 4, 7 and 14 of number 28 sum up to 28.

- A *pair of amicable numbers* is a pair of numbers such that the proper divisors of one number sum to the other number and vice versa.

- A *prime number p* is a positive integer that is exactly divisible by 1 and by itself only [2]. The numbers 1, 2, 3, 5, 7, 11, 13, 17, 19, 23, 29, 31, 37 and so on are prime numbers; the numbers 1, 2 are excluded because 1 is the very first integer of all positive numbers and 2 is the only even prime. All prime numbers, at the exception of 2, are integer odd; an even number cannot be prime since it would be divisible by at minimum 1, 2 and itself.

Euclid [3] provided a proof of the *Fundamental Theorem of Arithmetic*, postulating that every integer can be written as a product of primes in an essentially unique way. He also postulated that if the number $2^n - 1$ is prime, then the number $2^{n-1}(2^n - 1)$ is a perfect number.

In addition, Euclid's second theorem postulates that there are an infinite number of primes. This postulate triggered a quest among many notable mathematicians (Euler, Fermat, Legendre, Gauss, and others) to find the largest prime number ever, thus advancing prime number theory.

An estimate of the prime numbers within a range of numbers $(2, x)$ is found from $p(x) \sim x/\ln x$ (ln is the natural logarithm, or logarithm to the base e). For example, in the range 2 to 40 there are $40/\ln 40 = 40/3.688 = 10.843$ or ~ 11 prime numbers.

In addition, the nth random number p_n lies within the numbers $n/\ln n < p_n < n$ $[\ln(n) + \ln(\ln(n))]$, for $n > 5$. Thus, for $n = 10$, $10/\ln 10 < p_{10} < 10[\ln 10 + \ln(\ln 10)]$ or $4,3429 < p_{10} < 10[2.3025 + 0.8340] = 31.365$; the 10th prime number is 31 (not counting 1 and 2).

It was believed that finding the largest prime number and also the product of two very large prime numbers are calculation challenges; with supercomputing, these challenges are no more. The largest prime number by the year 2005 had 7,816,230 digits and the product of very large primes is relatively easily calculated. Yet, to this day, the most challenging problem is to determine the prime factors of a given very large integer, known as *factoring prime numbers*. In cryptography, the difficulty of *factoring prime numbers* from a very large integer has value because the prime numbers may serve as cryptographic keys or the seed of the keys. A *brute force* method for a given integer is the *direct search factorization*. This method tests systematically by trial division all possible factors that divide the given integer—a very tedious method and perhaps unsuccessful for extremely large numbers. That is, privacy is based on the premise that a third unauthorized party does not have the calculation capability to factor the same very large integer! Thus, a current concern to many is the development of a smart algo-rithm that efficiently and quickly factors large prime numbers, which will bring an end to cryptography that is based on the difficulty of *factoring prime numbers*. In his book *The Road Ahead* Bill Gates pointed out that the development of an easy way to *factor large prime numbers* that defeats the cryptographic system could be a disaster [4].

A category of prime numbers is the *twin primes*. Two prime numbers are called *twins* if their distance is 2, that is, $p_1 = p$ and $p_2 = p + 2$. For example, the primes 3 and 5 are twins, and so are the prime numbers pairs (5, 7), (11, 13), (17, 19), (29, 31), (41, 43) and so on [5].

2.4 MODULUS ARITHMETIC

The division of two rational numbers α and β is written as:

$$\alpha = \beta x Q + R,$$

where x denotes *multiplication*, α is the divident, β the divisor, Q is the quotient, and R the remainder.

Modulus arithmetic uses a notation such as $R = (\alpha)\text{mod}\beta$ or simply $R = \alpha\text{mod}\beta$, denoting that α is divided by β leaving a remainder R; the divisor β in this case is called the *modulus*.

For example, $1 = 13 \ (\text{mod}3)$ means $13 = 3 \times 4 + 1$, or that 13 is divided by 3 leaving a remainder of 1.

Modulus arithmetic has certain interesting properties with excellent applicability to cryptography and to Shift Cipher method. The following properties are common in cryptography:

- $(g\text{mod}\alpha)^x = (g^x)\text{mod}\alpha$
- $(g^x\text{mod}\alpha)^y\text{mod}\alpha = g^{xy}\text{mod}\alpha$
- $g^{xy}\text{mod}\alpha = g^{yx}\text{mod}\alpha$

In addition,

- $R = 0$ when α is exact multiple of β, and
- Two numbers A and B are equal if after division by m they yield the same remainder; that is: $A = B$ if $A(\text{mod}m) = B \ (\text{mod}m)$; this notation is often simplified to $A = B \ (\text{mod } m)$.

The latter property is important because large numbers can be reduced using mod$m$ operations. For example, consider the number $11 \times 17 = 187$. Then, using mod12 arithmetic there is:

$$187 = 15 \times 12 + 7, 7 = 187(\text{mod }12), \text{ or } 11 \times 17 = 7.$$

The latter example reveals another implication: calculating the product 11×17 is fast. The reverse however, that is given the product 187, finding the two numbers that produced it is not as fast, and it is extremely difficult if the two primes are very large.

2.5 GREATEST COMMON DIVISOR

Many cryptographic algorithms need to compute the greatest common divisor (*gcd*) of two positive numbers. Because a precision *gcd* computation is tedious, complex, time consuming and expensive, algorithms have been devised to compute *gcd* with fewer computational steps and accuracy commensurate with the algorithm. Among these algorithms are the *classical Euclidean algorithm*, the *non-classical binary algorithm*, the *Lehmer gcd algorithm* and the *binary extended Euclidean algorithm*.

A consequence between prime numbers and integers is: the *gcd* between a prime and an integer is 1.

In this section we describe the classical Euclidean *gcd* algorithm and the binary *gcd* algorithm.

2.5.1 The Classical Euclidean Algorithm

The *Euclidean algorithm* for computing the *gcd* of two integer numbers a and b with $a > b$, is provided executing the following steps:

1. While $b > 0$, calculate $a \bmod b$,
2. move b to a,
3. move the calculated value $a \bmod b$ to b.
4. Return a

A numerical execution of the Euclidean algorithm for $a = 2136$ and $b = 936$ yields a $gcd = 24$, as shown in Table 2.1.

Where a is the dividend, b the divisor, Q is the quotient and R is the remainder, $(R = a \bmod b)$.

The *extended Euclidean algorithm* finds the *gcd* of two positive integers A and B by finding two unique integers a and b such that $a \times A + b \times B = gcd(A, B)$.

2.5.2 The Binary GCD Algorithm

The *binary gcd algorithm* entails repeated division by 2. In binary logic, repeated division by 2 is accomplished by shifting a binary number in a shift register, and it is described with the following algorithmic example.

For two a and b positive numbers, $a > b$, and three registers, a, b and g, the binary division is accomplished algorithmically as follows:

1. Preset the two shift registers with the values of a and b, and initialize the divisor register g to 1.
2. For as long as $a > 0$ and $b > 0$:
   2.1. For as long as both a and b are even, keep dividing a and b by 2, and keep the latest divisor value in register g; that is, if there are n iterations the value in g is $g = 2^n$.
   2.2. For as long as a is even but not b, keep dividing a by 2, else for as long as b is even, keep dividing b by 2.

Table 2.1. Calculating the *gcd* for $a = 2136$ and $b = 936$ using the Euclidean algorithm

	a	$Q \times b$	$R = a \bmod b$
1.		$2136 = 2 \times 936 + 264$	
2.		$936 = 3 \times 264 + 144$	
3.		$264 = 1 \times 144 + 120$	
4.		$144 = 1 \times 120 + 24$	
5.		$120 = 5 \times 24 + 0$	
6.	**24**	—	—

Table 2.2. Calculating the *gcd* for the two numbers $a = 2136$ and $b = 936$

i:		1	2	3	4	5	6	7	8	9	10
a:	2136	1068	534	267	75	75	27	3	3	3	**3**
b:	936	117	117	117	117	21	21	21	9	3	**0**
g:	1	$2^1 = 2$	$2^2 = 4$	$2^3 = 8$	8	8	8	8	8	8	8
gxt:	—	—	—	—	—	—	—	—	—	—	$3 \times 8 = 24$

2.3. If both a and b are odd, then place the value $|a - b|/2$ in a temporary register t.

2.4. If $a \geq b$, then move the contents of t into a, else move t into b.

2.5. Return the product *gxt* to step 2

3. If $a = 0$ or $b = 0$, the *gcd* = *gxt*.

A numerical problem using the binary algorithm for $a = 2136$ and $b = 936$ yields *gcd* = 24 as follows (notice that a and b remains odd after the third division), (Table 2.2)

That is, the calculated *gcd* is 24.

2.6 GROUPS

A group $G(*)$ consists of a *set of elements* (or numbers) with a custom-defined binary operation $(*)$ on G such that any two group elements a and $b,$ a third element $c = a * b$ is defined that belongs to the group G. Note, some prefer the notation (#) instead of $(*)$ to denote an abstract operation in groups; here, we have no preference and use them interchangeably.

A group G satisfies the three abstract axioms:

• The *associative operation* $a * (b * c) = (a * b) * c$ for all $(a,\ b,\ c)$ in G holds.

• The group G has an *identity element* I such that $a*I = I*a = a$ for all elements a in G

• There is an *inverse of a* in G, a^{-1}, such that $a * a^{-1} = a^{-1} * a = I$.

Furthermore:

• The group G is abelian (commutative) if $a * b = b * a$.

• The group G is finite if $|G|$ is finite,

• *Order of group* is the number of elements in the finite group G.

EXAMPLE 2.1: *Consider the finite group of all integers Z_n with elements from 0 to n − 1 (including negative numbers), and basic* *operation the traditional addition (+) instead of an abstract operation:*

Then,

The associative operation $1 + (2 + 3) = (1 + 2) + 3$ holds,

The identity element is 0, since $0 + a = a + 0$

The inverse of any integer a in the group Z_n is defined as $-a$, such that $a + (-a) = (-a) + a = 0$

EXAMPLE 2.2: *Consider the set of positive integers M with basic operation the traditional multiplication (x) instead of an abstract operation.*

Then,

The associative operation $1 \times (2 \times 3) = (1 \times 2) \times 3$ holds,

The identity element is 1, since $1 \times a = a \times 1$

The inverse of any integer a in the group M is defined as a^{-1}, such that $a \times a^{-1} = a^{-1} \times a = 1$

Similarly, all *integers mod n*, where $n > 0$, form a group with identity 0 and the inverse of a as $(n - a)$.

Because groups can use custom-defined arithmetic rules, it make the solution of group problems difficult and thus groups are suitable in cryptography.

Two categories of groups are used in cryptography; the group Z_n defines the *additive group* of *integers mod n* (n is an integer); and the group Z_p defines the *multiplicative group* of *integers mod p* (p is a prime).

Group Z_n (with n elements) defines the traditional addition (+) which ends by reducing the result mod n. Doing so, the (addition result)mod n always yields numbers in the range 0 to $n - 1$. In addition, each element a in an additive group has an additive inverse element $-a$ such that $a + (-a) = 0$.

EXAMPLE 2.3: *Consider the group Z_{19}. Then*

$(13 + 17) \bmod 19 = 30 \bmod 19 = 11$, and in simplified Z_{19} reduced notation $30 = 11$

$(5 + 14) \bmod 19 = 19 \bmod 19 = 0$, and in simplified Z_{19} reduced notation $15 = 0$

Finally, the additive inverse of 5 is -5, in which case $-5 = 14 \bmod 19$ since $(5 + 14) \bmod 19 = 0$ or $[5 + (-5) + 14] \bmod 19 = 14 \bmod 19 = -5$.

Group Z_p (with integers from 1 to $p - 1$, where p is a prime) defines as basic operation the traditional multiplication x which ends by reducing the result mod n, thus assuring closure. The (product)mod n always yields numbers in the range 1 to $p - 1$. In addition, each element a in a multiplicative group has an additive inverse element b^{-1} such that in Z_p $a \times (b^{-1}) = 1$ or in full notation as $a \times (b^{-1}) \bmod n = 1$.

EXAMPLE 2.4:   *Consider the group Z_{19}. Then*

$$(5 \times 17) \bmod 19 = 85 \bmod 19 = 9, \text{ or in reduced } Z_{19} \text{ notation } 85 = 9$$

Similarly, $(15 \times 18) \bmod 19 = 270 \bmod 19 = 4$, or in Z_{19} $270 = 4$

The multiplicative inverse of 17 is 9^{-1}, in which case $9^{-1} = 17 \bmod 19$ since (9×17) $\bmod 19 = 153 \bmod 19 = 1$.

2.7   RINGS

A Ring $(R, +, x)$ consists of a *set of elements* R with two binary operations, $+$ for addition and x for multiplication on R that satisfy the axioms:

- $(R, +)$ is an abelian group.
- The operation x is associative: $a \times (b \times c) = (a \times b) \times c$ for all (a, b, c) in R
- There is a non-zero identity so that $1 \times a = a \times 1 = a$ for all a in R
- The operation x is distributive over the operation $+$, such that $a \times (b + c) = (a \times b) + (a \times c)$ for all (a, b, c) in R
- The ring is commutative is $a \times b = b \times a$

EXAMPLE 2.5:   *the set of integers with the traditional addition and multiplication comply with the above axioms and is a commutative ring.*

EXAMPLE 2.6:   *the set of integers with the addition and multiplication modulo* **n** *is a commutative ring.*

An elements b is an *invertible element* of a if $a \times b = 1$, where both a and b are elements of the ring R.

2.8   FIELDS

A field F is a commutative ring in which all non-zero elements have multiplicative inverses. The characteristic of a field is 0 if $1 + 1 + 1 \ldots + 1$ (m times) $\neq 0$, for $m \geq 1$.

EXAMPLE 2.7:   *The rational numbers* **Q**, *the real numbers* **R**, *and the complex numbers* **C** *comply with the aforementioned definitions and thus they form fields. Conversely, the ring of integers in Example 2.6 above does not form a field since the only non-zero elements with multiplicative inverses are* **1** *and* **−1**.

A field is finite if it contains a finite number of elements. The *order of F* is the number of elements in *F*.

If the finite field *F* contains p^m elements for some prime *p* and integer $m \geq 1$, then for every prime power order p^m, there is a unique finite field of order p^m. This field is denoted F_{pm}, or $GF(p^m)$; this notation is used in elliptic curve cryptography.

The finite field F_p where *p* is a prime number, consists of the numbers from 0 to $p - 1$ and the basic operations are defined to be the additive and multiplicative and all calculations end with the reduction modulo *p*; operations such as division, subtraction and exponentiation are derived in terms of addition and multiplication. In this case, all non-zero elements have a multiplicative inverse.

EXAMPLE 2.8: *In F_{23}:*

$$(10 \times 4 - 12) \bmod 23 = 28 \bmod 23 = 5$$

$7 \times 10 \bmod e23 = 70 \bmod e23 = 1$, thus 7 and 10 are inverse of each other:

$$7^{-1} \bmod e \, 23 = 10 \text{ and } 10^{-1} \bmod e \, 23 = 7$$

$$\left(9^3/10\right) \bmod 23 = (729/10) \bmod 23 = (16/10) \bmod 23 = (16 \times 7)$$
$$\bmod 23 = 272 \bmod 23 = 20.$$

Thus, if *p* is a prime number (3, 5, 7, 11, 13, …), the *integer modp*, denoted *Zp*, is an additive group followed by mod*p*, or a multiplicative group followed by mod*p* (in this case the 0 elements are not included) have an identity 1.

2.9 THE FERMAT'S THEOREM

The Fermat's theorem states that if *p* is a prime number and *a* is a number such that $0 < a < p$, then:

$$a^{p-1} \bmod p = 1$$

As an example, consider the prime number $p = 7$. Then, the last column in Table 2.3 confirms the result of Fermat's theorem.

Table 2.3. Calculation for $p = 7$ and $0 < a < 7$ according to Fermat's theorem

$P = 7$	a	$a^2 \bmod 7$	$a^3 \bmod 7$	$a^4 \bmod 7$	$a^5 \bmod 7$	$a^6 \bmod 7$
	2	4	1	2	4	1
	3	2	6	4	5	1
	4	2	1	11	2	1
	5	4	6	2	3	1
	6	1	6	1	6	1

2.10   THE EULER'S THEOREM

The Euler's theorem states that if n and a are positive integers such that $a < n$, then:

$$a^{\Phi(n)} \bmod n = 1$$

where $\Phi(n)$ is the Euler's *phi function*:

$$\Phi(n) = n(1 - 1/p_1) \ldots (1 - 1/p_m)$$

where $p_1, \ldots, p_m$ are the prime numbers that divide evenly into n, including n if it is a prime.

If n is a prime number, then $\Phi(n) = n(1 - 1/n) = n[(n - 1)/n] = (1 - n)$ and thus Euler's theorem becomes:

$$a^{n-1} \bmod n = 1$$

That is, Euler's theorem is a special case of Fermat's theorem.

Another special case is when the modulus is the product of two prime numbers p and q, $n = pq$. Then, $\Phi(n) = n(1 - 1/p)(1 - 1/q) = (p - 1)(q - 1)$.

For example, consider for simplicity the two smallest prime numbers $p = 3$ and $q = 5$. Then $\Phi(15) = 15(1 - 1/3)(1 - 1/5) = (3 - 1)(5 - 1) = 2 \times 4 = 8$. Then $a^8 = \bmod 15 = 1$ for values of a that have no common divisors with 15, such as 2, 4, 7, 8, 11, 13, 14 (Table 2.4).

Table 2.4. Calculation for the Euler's special case for $p = 3$ and $q = 5$

$n = (3 \times 5) = 15$	a	$a^8 \bmod 15$	Common divisor
15	2	1	—
15	3	6	3
15	4	1	—
15	5	10	5
15	6	6	3
15	7	1	—
15	8	1	—
15	9	6	3
15	10	10	5
15	11	1	—
15	12	6	3
15	13	1	—
15	14	1	—

2.11 EXCLUSIVE-OR

Computer generated text as well as digitized analog signals, such as voice and video, are mixed with a cryptographic key to generate the cipher text using the logic Exclusive-Or (XOR) function. The output of a binary XOR function with two inputs, represented by the symbol $\otimes$, is defined in Table 2.5.

Binary XORs have some interesting properties:

$$A \otimes A = 0$$

$$A \otimes 0 = A$$

$$A \otimes 1 = {\sim} A; (\sim\! A \text{ denotes the inverse of } A)$$

$$A \otimes B = B \otimes A$$

$$A \otimes B \otimes C = A \otimes C \otimes B = B \otimes C \otimes A$$

If $A \otimes B = C$, then $A \otimes C = B$ and $B \otimes C = A$;
(this is used for ciphering and deciphering)

Based on the first two XOR properties, it follows (easily provable) that:

$$A \otimes A \otimes A = A, (\text{odd number of } As), \text{ and}$$

$$A \otimes A \otimes A \otimes A = 0, (\text{even number of } As), \text{ or in general}$$

$nA\otimes = 0$ for $n =$ even, and $= A$ for $n =$ odd (this is used to calculate the parity).

In cryptographic systems, data is XOR'ed with the cipher key bit by bit, byte by byte or block by block of data. Bit by bit is accomplished by synchronizing serial data stream with a serial digital key, typically at the serial output of a transceiver. Byte by byte is accomplished at the data bus at the register level (8, 16, 32 or 64 bit) and prior to being serialized. Block by block is similar to byte by byte, although a block is an integer multiple of bytes. Based on this, XOR'ing data with the cipher key produces the cipher text:

Plain text: **11001101|01001101|01011101|**...
Cipher key: **01110100|11110110|00011011|**...
Cipher text: **10111001|10111011|01000110|**...

Table 2.5. XOR logic function

A	B	$A \otimes B$
0	**0**	**0**
0	**1**	**1**
1	**0**	**1**
1	**1**	**0**

Based on the property, if $C = A \otimes B$ then $A = B \otimes C$, and using the same cipher key on the cipher text, the original text is recovered:

Cipher text:	**10111001|10111011|01000110|**...
Cipher key:	**01110100|11110110|00011011|**...
Plain text:	**11001101|01001101|01011101|**...

2.12 RANDOM NUMBERS

Here we provide a qualitative definition of random numbers.

Consider a number N that consists of a very long sequence of numeric symbols or digits. Consider a sliding window of a width in terms of a sequence of digits, n, where $n > 2$; call n a contiguous sequence of numbers "seq." Assume that the sliding window is set at a sequence width, it scans the number N and it counts the occurrences of the sequence "seq," if it finds any. The sequence length is variable and the scanning process repeats for a larger width, and so on. If the window finds no sequence of any length that repeat in the number N, then the number is called a *random number.*

Now, assume that the sliding window finds a sequence of some length L that repeats itself more than once. Then, if within L there is no repetitive sub-sequence, then the sequence of digits L is a random number of finite length or a *pseudo-random number.*

Random numbers are found in nature. For example, the number $\pi = 3.14 \ldots$ after the decimal point is random. Random numbers of finite length can be generated with simple circuitry known as *pseudo-number generators* (PRNG). Because these RNGs are also deterministic, the generated sequence is always known; therefore, they are also called *deterministic random number generators* (DRNG).

In cryptography, binary random numbers are typically used to generate cipher keys. In this case, binary random numbers are generated with digital circuitry (typically, a shift register with feedback via XORs implementing a binary polynomial) called *random number generators* (RNG) and a seed value (also known as the initial value (IV)) from which the RNG starts generating a random number.

In cryptographic applications, a useful RNG should produce a stream such that the next bit in the random stream is forward unpredictable if the seed is unknown. Forward unpredictability depends on the RNG selection and the selected seed.

Unpredictability is also backward; that is, the seed cannot be determined from the generated values in the stream. That is, given the random stream the seed should not be able to be derived from. Also, there should be no correlation between a seed and any value generated from that seed.

The generated RN stream must satisfy the following:

- **Uniformity:** the number of 1s and 0s in the sequence is equal; that is, the probability of generating a 1 or a 0 as the next digit is 0.5.
- **Consistency:** the RNG must consistently generate the same sequence for each seed (initial value of the register-based RNG).

- **Scalability:** This is related to the sequence of the scanning window. If a sequence of numbers in the stream is truly random, any sub-sequence of numbers within it is random. Clearly in cryptography, the longer the random sequence, the better.

In general, random number generators are important in telecommunications; RNs are also used to scramble the signals at the physical layer to assure that the distribution of 1s and 0s in the transmitted signal has equal probability of occurrence and to also remove long strings of 0s and 1s. In cryptography, RNs are equally important and standard documents describe aspects of RNGs [6, 7]. See also Chapter 9 for RN testing.

REFERENCES

1. R.K. GUY, *Unsolved Problems in Number Theory, 3rd ed.* New York: Springer-Verlag, 2004.
2. R. CRANDALL and C. POMERANCE, *Prime Numbers.* Springer-Verlag, New York, 2001.
3. Euclid, Book IX, Elements.
4. BILL GATES, *Road Ahead*, 1995, p. 265, Viking Publishers.
5. P. RIBENBOIM, "Twin Primes." §4.3 in *The New Book of Prime Number Records.* New York: Springer-Verlag, pp. 259–265, 1996.
6. NIST Special Publication 800-22, *A Statistical Test Suite for Random and Pseudorandom Number Generators for Cryptographic Applications*, May 15, 2001.
7. ANSI X9.82, *Deterministic Random Bit Generators*, 1982.

Chapter 3

Ciphers and Algorithms

Cryptography implies that a plain text has been altered by the rightful sender, the source, such that it is either unintelligible or intelligible but with different meaning; this is the product of "mixing" the plain text with a cipher key, with a cipher text, or with a double meaning text (Figure 3.1). The original plain text can only be recovered from the cipher text by the rightful recipient using a deciphering method that, in its simplest form, can be the same cipher key or the same cipher text.

The benefit of the unintelligible text over the altered double-meaning text is that the former is generated using a provable cipher key or text, whereas the latter uses a pre-agreed phrase, the secrecy of which depends on trust and confidentiality. Because machine-generated provable cipher keys remove the human factor from the equation of secrecy, they are more popular, and attention is given to discovering cipher keys that are unbreakable and for which the key distribution method is secure. In that respect, different cryptographic methods have been devised using symmetric or asymmetric keys, or other more advanced methods.

3.1 SYMMETRIC/ASYMMETRIC CIPHERS

Cryptography based on machine-generated provable cipher keys fall in two major categories, *symmetric key cryptography* and *asymmetric key cryptography*.

Symmetric key cryptography uses the same key for ciphering plain text and deciphering cipher text at both end of the transmission path. In this case, encryption algorithms are based on random numbers, prime numbers and common divisors.

Asymmetric key cryptography uses one key for ciphering plain text and another key for deciphering cipher text. However, the two keys are mathematically interrelated. In this case, encryption algorithms are based on prime numbers, elliptic curves and other difficult mathematics.

This chapter describes ciphering algorithms according to either symmetric or asymmetric key cryptography.

Security of Information and Communication Networks, by Stamatios V. Kartalopoulos

Original plain text:	When market is low buy 2 million shares.
Unintelligible text:	GCVK KLTOIH OW MVA ASQ @ BXASZC FGRESL
Double meaning text:	When apples are red, send 2 truckloads; or,
	In the fall, 2 million birds move south.

Figure 3.1.   Cipher text: A. Unintelligible, and B. intelligible but not having the same meaning with the original text.

Table 3.1.   Characters of the alphabet are placed in cells, which are then numbered randomly

A	B	C	D	E	F	G	H	I	J	K	L	M	N	O	P	Q	R	S	T	U	V	W	X	Y	Z
1	2	3	4	5	6	7	8	9	10	11	12	13	14	15	16	17	18	19	20	21	22	23	24	25	26
5	9	21	15	1	25	14	22	2	26	16	17	18	7	4	11	12	13	23	24	3	8	19	20	6	10

3.2   SYMMETRIC CIPHERS

3.2.1   The Direct Substitution or Random Shift Cipher

The direct substitution shift cipher is a symmetric key cryptographic method that uses look-up tables, and thus it is fast.

Consider the 26 characters of the Latin alphabet, each placed in a cell numbered linearly from 1 to 26. Each cell is also associated with a a random order number of pair-wise reversible numbers from 1 to 26 (Table 3.1). Pair-wise reversibility is defined by pairs of numbers as in $1 \rightarrow 5$ and $5 \rightarrow 1$ or $2 \rightarrow 9$ and $9 \rightarrow 2$. Notice that pair-wise reversible numbers may be defined in several different ways (for example, $1 \rightarrow 3$ and $3 \rightarrow 1$, $1 \rightarrow 12$ and $12 \rightarrow 1$, and so on).

For a given plain text, replace each character in it by its corresponding random number, and then by a character that corresponds to the linear number. Doing so, the plain text "GOOD DAY" is ciphered to "NDDO OEF" following the steps in Table 3.2.

To decipher the cipher text, the reverse process is followed. The cipher text NDDO OEF is converted to its linear numbers, which are associated with the same random order of pair-wise reversible numbers, and then the plain text is recovered (Table 3.3).

The direct substitution shift cipher cryptographic method maintains the frequency of occurrence (although of different characters), which is a weakness that can be exploited by cryptanalytic methods and thus find the key if the text is lengthy.

To remove the frequency of occurrence, a random sequence of length $nx26$ may be used, the alphabet in Table 2.1 is repeated n times and each character in the n alphabets is associated with a corresponding number of the random sequence. Thus, the same letter is not associated with the same number (only by accident) in moving from alphabet to the next alphabet.

Table 3.2. Converting plain text in cipher text with the direct random shift method

G	O	O	D	D	A	Y	Plain text
7	15	15	4	4	1	25	**Linearly corresponding numbers**
14	4	4	15	15	5	6	**Pair-wise reversible numbers**
N	D	D	O	O	E	F	**Cipher text**

Table 3.3. Recovering the plain text with the direct random shift method

N	D	D	O	O	E	F	Cipher text
14	4	4	15	15	5	6	**Linearly corresponding numbers**
7	15	15	4	4	1	25	**Pair-wise reversible numbers**
G	O	O	D	D	A	Y	**Plain text**

Table 3.4. Characters of the alphabet are placed in numbered cells

A	B	C	D	E	F	G	H	I	J	K	L	M	N	O	P	Q	R	S	T	U	V	W	X	Y	Z
1	2	3	4	5	6	7	8	9	10	11	12	13	14	15	16	17	18	19	20	21	22	23	24	25	26

This method requires synchronization at the receiver, which, however, is not an insurmountable issue.

3.2.2 Shift Cipher

The Shift Cipher is a symmetric key cryptographic method and it uses extensively modular arithmetic (see Chapter 2). In modular arithmetic, the division of two rational numbers α and β is written $\alpha = \beta \times Q + R$, where x denotes *multiplication*, α the dividend, β the divisor, Q the quotient, and R the remainder. Thus, $\alpha = 14$ and $\beta = 3$ yields $Q = 4$ and $R = 2$, written as $2 = 14 \bmod 3$.

The shift cipher cryptographic method is explained with the following example.

Assume the English alphabet with 26 characters, A to Z, each one is placed in a cell and each cell is associated with a number, $A => 1$ to $Z => 26$; Table 3.4 provides an example in ascending order:

Given a plain text, each character in it is replaced by its associated number, **N**, which is shifted by n cells, where $n < 26$ is the *shift key*, and then calculate the $n \bmod 26$. The cipher text is obtained by replacing the calculated numbers by the corresponding characters. Thus, for $n = 7$, the plain text "GOOD DAY" is ciphered to "LDDO ORF"; the steps involved are shown in Table 3.5.

Table 3.5.   Converting plain text in cipher text with the shift key method

G	O	O	D	D	A	Y	Plain text
7	15	15	4	4	1	25	**Corresponding numbers**
14	22	22	11	11	8	6	**Shifted by 7**
12	4	4	15	15	18	6	**mod26**
L	**D**	**D**	**O**	**O**	**R**	**F**	**Cipher text**

To retrieve the original plain text from the cipher text, one proceeds backwards converting the cipher text to corresponding numbers in the alphabet, subtract *7mod26* and convert the sequence of produced numbers to a sequence of characters in the alphabet.

The shifted cryptographic method still maintains the character frequency of occurrence, which is a cryptographic weakness making cryptanalysis simpler.

3.2.3 The Permutation Cipher

The Permutation cipher is a symmetric key cryptographic method. This method does not replace each character of the plain text by another, but It rearranges the characters of the same text in a specified random-like order using a permutation algorithm. Thus the plain text,

THE SKY IS BLUE

may be ciphered, based on a spatial permutation algorithm, to

EKS TIU TB LHES.

Changing the algorithm yields a different permutation of characters in the text, which makes the cryptographic algorithm more difficult and thus stronger. However, because the frequency of characters in the text remains the same, it is possible to find the permutation algorithm if the text has either a sufficiently long length.

3.2.4 Block Ciphers

In cryptography, besides the aforementioned single character or bit manipulations that can be encoded and decoded bit at a time and relatively easily, one may also consider a combination of mathematical operations such as permutations, modulus arithmetic, re-arrangements, ciphering and more in order to add to the encryption complexity. Moreover, one may consider encrypting a plain text of characters or bits that have partitioned in blocks of characters or bits. In this case, the encryption method is classified as *block cipher* [1]. The first block cipher using complex and convoluted methods with 128-bit keys was developed by IBM in 1970 and it was code named *Lucifer*.

Typically, blocks are numbered from left to right and bits in them also from left to right with the left most bit numbered one. Block ciphers operate in several modes, four of which are the most popular:

- *The* **electronic codebook** *(ECB) mode* uses a secret key to encrypt the plain text block to a cipher text block. As such, two identical plaintext blocks produce the same cipher text.

- *The* **cipherblock chaining** *(CBC) mode* exclusive-Ores the previous cipher text with the current plain text prior to encryption. As such, two identical plain text blocks do not produce the same cipher text.

- *The* **cipher feedback** *(CFB) mode* allows data to be encrypted in units smaller than the block size, and it is a self-synchronizing stream cipher; self-synchronizing ciphers calculate each bit in the stream as a function of the previous *n* bits in the stream. This process remains synchronized with the encryption but it propagates errors.

- *The* **output feedback** *(OFB) mode* uses an internal feedback mechanism that prevents the same plaintext block from generating the same cipher text block. The OFB is similar to synchronous stream ciphers; synchronous stream ciphers use the same keystream generation function at the transmitter and receiver. Although synchronous stream ciphers do not propagate errors, they are by nature repetitive.

3.2.5   The Data Encryption Standard (DES)

The block cipher *Lucipher* was adopted in 1977 by the National Institute of Standards and Technology (NIST) and it was renamed *Data Encryption Standard* (DES) which partitions a binary message of "1"s and "0"s in 64-bit blocks [2].

The DES is a symmetric encryption algorithm. It uses a block cipher with a 56-bit key and an 8-byte block (to a total of 64 bits); the key is organized in eight 7-bit bytes and a bit is added to each byte adjusting its parity to odd and thus the added 8 bits acts as a key error detection and correction (EDC) mechanism.

The 56-bit key is used to transpose 64 bits. The transposed block is then permuted using another 56-bit key which is derived from the initial key. The result of this step undergoes 16 more encryption steps, each using 56-bit keys that are derived from the initial key. The result of this undergoes a swap operation and then a final transposition.

Each of the 16 encryption steps consists of a very complex process that starts with partitioning the 64-bit blocks in two halves, each of 32 bits. The first half, A32, remains as is whereas the second half, B32, is expanded to 48 bits by transposing some bits and duplicating others. The expanded second half is XOR ciphered with a 56-bit key. The ciphered 48 bits are then subdivided into eight 6-bit groups. Each 6-bit group by substitution produces a 4-bit result and all eight reduced groups form a 32-bit string. Bits in the string are transposed and the result C32 is XORed with the first 32-bit half, A32, from the initial partitioning. Now, the result of this and B32 are combined to form a 64-bit encrypted block. Enciphering proceeds from the left most bit.

In another version, this cryptographic method uses three independent keys for each block of data; in this case, this is known as the Triple Data Encryption Algorithm (TDEA) and it is specified to protect sensitive information

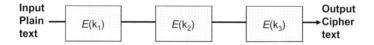

Figure 3.2.   Triple DES Encryption in ECB mode.

61	17	40	9	45	15	63	5
36	51	2	34	20	58	27	39
47	11	30	46	8	23	38	14
22	59	19	26	57	61	64	6
52	48	35	54	33	1	28	44
31	3	41	12	50	18	60	29
42	53	24	56	4	37	55	13
7	25	16	21	32	43	10	49

Figure 3.3.   The first step in DES encryption is a permutation of all 64 bits in a block.

of Federal organizations (TDEA is described in the same standard with DES). In this case, TDEA uses a 168-bit key and provides the equivalent of 112 bits of security.

The DES and TDEA standard [3] describes for modes: the Electronic CodeBook (ECB), the Cipher Block Chaining (CBC), the Cipher Feedback (CFB), and the Output Feedback (OFB). The triple key in the DCB mode (per FIPS PUB 46-3) (Figure 3.2).

Enciphering a block with the DES method is described in three major steps:

1. It starts with an initial permutation (**IP**) of the 64-bit block; in short, it organizes all 64 bits of the block in an 8×8 matrix and reshuffles all bits in the matrix (Figure 3.3). The permuted matrix is then divided in two halves, the left (**L**) and the right (**R**).

2. This process continues with a sequence of complex computations and permutations using 16 keys defined by a cipher function f and a key schedule **KS**. The function f is given in terms of primitive functions called the selection functions (S_i) and the permutation function P, Figure 3.4.

3. It finishes the outcome of step 2 with the inverse of the initial permutation (IP^{-1}) to reshuffle the results one more time (Figure 3.5).

Although the DES cryptographic method seems complex, convoluted and confusing (and thus seemingly difficult to break) in 1998, the Electronic Frontier

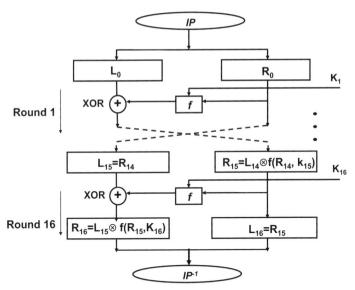

Figure 3.4. The DES entails a complex process of several functions and keys.

30	7	14	51	39	20	45	10
38	17	57	29	2	33	52	26
1	53	21	60	25	43	13	37
28	11	58	19	63	5	61	44
42	50	6	56	34	24	59	16
8	49	40	12	62	64	32	47
54	18	31	46	22	3	55	36
27	41	4	23	48	35	15	9

Figure 3.5. The third step in DES encryption reshuffles the outcome of step 2.

Foundation built a computer-based *DES cracker* that was able to decipher DES encrypted messages [4, 5].

3.2.6 The Advanced Encryption Standard (AES)

In view of DES's demise, in 1997 NIST issued a request for proposals (RFP) for an *Advanced Encryption Standard* (AES). From 15 proposals submitted, in 2000 NIST announced that the algorithm submitted by Vincent Rijmen and Joan Daemen, called the *Rijndael algorithm* [6], was the chosen one [7]. That is, the Rijndael is a federal government approved encryption algorithm defined in Federal Information Procedure Standard (FIPS) no. 197 (2001).

The Rijndael algorithm is a symmetric block cipher algorithm, it is mathematically complex and it is based on several complex mathematical concepts such as field theory, Galois fields (GF(2^8)), irreducible polynomials, equivalence classes, and more [8–10].

The Rijndael algorithm processes blocks of 128 bits (the input and output of this algorithm is 128 binary bits) and it specifies three keys, 128, 192, and 256 bits long. The bits are numbered starting from 0 and end at $n - 1$ (where $n = 128$, 192, 256) and thus a bit index i has a range $0 \leq i < 128$, $0 \leq i < 192$, or $0 \leq i < 256$; NSA has approved the 128-bit AES for use up to SECRET level documents, and the 192-bit AES for up to TOP SECRET level.

An 8-bit byte is treated as a unit in the Rijndael algorithm; a byte is denoted as $\{b_7, b_6, \dots b_1, b_0\}$, where b_j has the binary value 0 or 1. Thus, the byte $\{10010101\}$ is written algebraically as $1x^7 + 0x^6 + 0x^5 + 1x^4 + 0x^3 + 1x^2 + 0x^1 + 1x^0$ or briefly $x^7 + x^4 + x^3 + x^0$, and in hexadecimal notation $h95$ or briefly 95; notice that the byte is partitioned in four high bits and in four low bits when converting to hexadecimal.

A string of bits is subdivided in groups of contiguous bytes to form arrays of bytes. Thus, a block contains 128 bit or 16 bytes and a key consists of $0 \leq n < 16$, $0 \leq n < 24$, or $0 \leq n < 32$ bytes.

The AES block is subdivided in a sequence of sub-blocks of 8-bit bytes organized in a two-dimensional matrix of bytes known as the *State*; a 128-bit block is organized in a 4-rows by 4-columns State, and a 192-bit block in a 4-rows by 6-columns State.

Then an iterative process starts that consists of *modulo-2 addition* (XOR), *multiplication in GF(2^8), byte substitution, row shifting, column mixing*, and *key addition*. After 10 rounds of this process, the end result is the final cipher text.

The proposed AES algorithm has been coded using 8-bit and 32-bit microprocessors. For a 128-bit key and 128-bit block, it takes 8390 cycles using an 8-bit microprocessor and a 2100 cycles using a 32-bit microprocessor.

The complete sequence and description is beyond the purpose of this book; it is the subject of the AES specification listed in aforementioned references. However, for completeness and for appreciating the complexity in cryptology, two mathematical operations are explained via examples, modulo-2 addition and multiplication in GF(2^8), which is used in AES as well as in other cryptographic methods.

EXAMPLE 3.1:   *Modulo-2 addition*

Addition is performed in modulo-2 (XOR):

In polynomial form:	$(x^7 + x^6 + x^3 + x^1 + 1) + (x^6 + x^1 + 1) = (x^7 + x^3)$
In binary form:	$\{1\,1\,0\,0\,1\,0\,1\,1\} \otimes \{0\,1\,0\,0\,0\,0\,1\,1\} = \{1\,0\,0\,0\,1\,0\,0\,0\}$
In hexadecimal form:	$\{C\ B\} \otimes \{4\ 3\} = \{8\ 4\}$

EXAMPLE 3.2:   *Multiplication in GF(2^8)*

Preliminaries: Multiplication in GF(2^8) is accomplished by calculating the multiplication of two polynomials modulo (*an irreducible polynomial of degree 8*).

An irreducible polynomial is one that is divided exactly by x^0 (that is, by 1) and by itself only (this definition is similar to prime numbers); for example, the polynomial $x^8 + x^4 + x^3 + x^1 + x^0$, in binary {01}00011011, and in hexadecimal {01}1B form is irreducible; here, the 9th bit (or x^8) is enoted as {01}.

Multiplication in GF(2^8): Now, the multiplication of the two binary polynomials 0101 0111 or {57} and 1000 0011 or {83} in GF(2^8) is achieved as follows.

First, the product {57} * {83} = {C1} is calculated in binary polynomial form similar to algebraic polynomials where exponents are added but the coefficients are not and remain always 1, as in $x^1 + x^1 = x^1$:

$$\left(x^6 + x^4 + x^2 + x^1 + x^0\right)*\left(x^7 + x^1 + x^0\right) = x^{13} + x^{11} + x^9 + x^8 + x^6 + x^5 + x^4 + x^3 + x^0$$

Then, the product in GF(2^8) is calculated by calculating the product modulo an irreducible polynomial, such as $x^8 + x^4 + x^3 + x^1 + x^0$:

$$x^{13} + x^{11} + x^9 + x^8 + x^6 + x^5 + x^4 + x^3 + x^0 \text{ modulo}$$
$$\left(x^8 + x^4 + x^3 + x^1 + x^0 = x^7 + x^6 + x^0\right),$$

Or in binary form:

$$(10101101111001) \text{ modulo } (100011011) = (11000001)$$

The result of multiplying in GF(2^8) produces an 8-bit byte. In AES, modular reduction to 8-bit bytes is needed for processing blocks that are multiple of 8-bit bytes.

Although this modular reduction seems complex, notice that the polynomial modulo operation is easily performed by direct binary division, which is performed step-by-step in Figure 3.6.

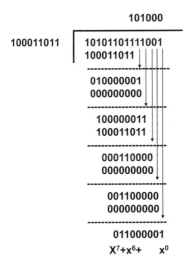

Figure 3.6.   Binary division of two polynomials yields the remainder which is the result in *GF(2^8)*.

3.2.7 The RC4 Algorithm

The Rivest Cipher version 4 (RC4) Algorithm, designed by Ron Rivest in 1987, is a symmetric cryptographic method for which a variable key-size stream cipher is used, which is generated by a *pseudo-number generator*.

According to it, an initial key (from 1 to 256 bytes) stored in a temporary vector T creates a variable 256-byte key in a state vector S; thus, the RC4 performs byte-wise operations and random permutations.

The state vector consists of 256 elements $S[0]$, $S[1]$, ... , $S[255]$. At initialization, the state elements $S[i]$, $i = 0$ to 255 contain one of the $2^8 = 256$ binary combinations in an ascending order, 0 (00000000) to 255 (11111111).

The temporary vector T also consists of 256 elements. At initialization, T is preset with the cipher key. If the key is 256-bytes long it fills all elements of T. If the key is k bytes (where $k < 256$), then k bytes fill the first k elements of T and the key repeats itself as many times as necessary to fill T.

The role of each T element, $T[i]$, is to cause a swap of element i in S, $S[i]$, with element j in S, $S[j]$, which is defined according to an algorithm. The swapping process is an iterative process and it starts from element $i = 0$, $S[0]$, and ends with element $i = 255$, $S[255]$.

To find the element j with which element i is swapped, one executes a *modulo256* calculation between $S[i]$, $T[i]$ and j. Thus, for element $i = 0$, element j_0 is calculated from $j_0 = \{S[0] + T[0]\}\bmod256$. Now, for element $i = 1$, the value of the next j, j_1, is calculated from $j_1 = \{j_0 + S[1] + T[j_0]\}\bmod256$. Similarly, for element $i = 2$, the value of the next j, j_2, is calculated from $j_2 = \{j_1$ $S[2] + T[j_1]\}\bmod256$, and so on, to the final element $j_{255} = \{j_{254}$ $S[255] + T[j_{254}]\}$ mod256.

Since the initial vector S is predefined (0 to 255), and the temporary vector T is also predefined (it contains the pseudo-random initial key), all swapping calculations can be done a priori and then use vector S in its final (swapped) state as the cipher key; this would require one memory to store the cipher key in S, eliminating T. Alternatively, if T is generated byte at a time, then since the initial S is known, all is needed is a one line iterative calculation, the result of which is stored in a register and which is the current element of the cipher key.

Because of the algorithm simplicity, RC4 has found acceptance in several protocols such as the WiFi Protected Access (WPA), the Wired Equivalent Privacy (WAP), and in Secure Sockets Layer/Transport Layer Security (SSL/TLS). RC4 was initially a RSA Security private algorithm, but an anonymous posted it on the Internet.

3.2.8 The RC5 and RC6 Algorithms

The RC5, designed by Ronald Rivest for RSA Data Security (now RSA Security) in 1994 [11], is a symmetric fast block cryptographic algorithm that has adopted many of the RC4 concepts (State vector, etc) but with added security and flexibility.

In addition, RC5 is designed to be implementable in different processor bus widths, 8-, 16-, 32- and 64-bit.

To support flexibility at all levels of security and efficiency, the RC5 algorithm has variable key size, variable block size, and variable number of rounds:

- The key is from 0 to 2040 bits (128 suggested). The key schedule is complex and it expands the key using binary expansions.

- The block size is 32 bits for experimentation and evaluation, 64 bits for direct DES replacement, and also 128 bits (64 suggested).

- The number of rounds ranges from 1 to 255 (12–20 suggested).

The RC5 algorithm performs three routines: key expansion, encryption, and decryption.

1. In key-expansion, the user-provided secret key is expanded to fill a key table the size of which depends on the number of defined rounds. The key table is used in both encryption and decryption, which constitutes it a symmetric algorithm.

2. The encryption routine consists of three primitive operations: integer addition, bitwise XOR, and variable data-dependent rotation.

   Thus, with the key-expansion done and the array $S[0, 1, t–1]$ computed, the input block is put in two registers A and B. Then the following pseudo-code calculates the encryption [pseudo-code adapted for illustrative purpose from [11]:

$$A = A + S[0]$$
$$B = B + S[1]$$

For $i = 1$ to r, do:

$$A = ((A \otimes B) << B) + S[2+i]$$
$$B = ((B \otimes A) << A) + S[2+i+1]$$

3. The decryption routine performs the reverse of the encryption:
   For $i = r$ to 1, do:

$$B = ((B - S[2+i+1]) >> A) \otimes A$$
$$A = ((A - S[2+i]) >> B) \otimes B$$
$$B = B - S[1]$$
$$A = A - S[0]$$

Although as shown RC5 is simple to code (few lines of code and thus short execution time), its security relies on the extensive use of data-dependent rotations, on a mix of different operations and on the lengthy blocks (the 12-round RC5 with 64-bit blocks has been susceptible to differential attack using 2^{44} chosen plaintexts. Typically, the use of rotations makes differential and linear cryptanalysis difficult.

The RC6 is a block cipher based on RC5 and it was designed by Rivest, Sidney, and Yin for RSA Security in it was submitted to NIST's RFP for AES. The RC6 is also an algorithm with variable block size, key size, and number of rounds and with a key upper limit of 2040 bits. RC6 has the two new features, integer multiplication and four $b/4$-bit working registers (two $b/2$-bit registers in RC5), where b is the block size.

- Integer multiplication decreases the number of rounds and increases the speed of the cipher.
- The four working registers allow for 32-bit operations and AES architecture compatibility.

3.2.9 Other Well-Known Algorithms

Chapter 1 described an ancient cryptographic method that has been known as "*skytale*" or *staff* (Figure 1.3). This method entails a ribbon of length L~$mx\pi D$, where m is a real positive number (typically >5), width W, and D is the diameter of the cylindrical staff (or $2R$). Based on this, to express the helicoidal or spiral form of the ribbon wrapped around the staff one needs to define the sense of the helix, $S = +1$ or -1, for counter-clockwise (also called *enantiomorphous or left-handed*) or clockwise (also called *right-handed*), and the step of helix is W (the width of the ribbon). Additionally, one needs to also consider the number of rows to be written on the surface of the staff after the ribbon is wrapped around it; four rows imply that a complete revolution is divided in four quadrants, and six rows in 60° sectors. Thus, if the aforementioned parameters are not known to an unauthorized entity, a moderately long message is difficult to decrypt (although not impossible).

Mathematically, the helix is a space curve described by the parametric equations [12]:

$$x = R\cos\Theta$$
$$y = R\sin\Theta$$
$$z = W\Theta$$

where Θ varies from 0 to 2π, R is the radius of the staff, and z is the direction of evolution of the helix.

The arc length S of the helix is given by

$$S = \sqrt{[(R^2 + z^2)\Theta]}$$

Cryptographically speaking, to make things more difficult this method is improved by forming the ribbon in a "*Moebius band*" and then wrapping it around the staff on the surface of which the message is written. In this case, although all encrypted characters of the message appear on the "*Moebius band*" or "*Moebius ribbon*" on a continuous surface, neither the start of the ribbon nor how the ribbon was folded and wrapped around the staff are known to an unauthorized interceptor. This improved method is called the "*Moebius skytale*". The Moebius ribbon was invented by mathematician A.F. Moebius (1790–1868) and it is also known as a one-sided surface and although it is simple to construct its mathematical complexity and properties have mystified scientists, philosophers and artists like M.C. Escher.

3.3 ASYMMETRIC CIPHER SYSTEMS

Symmetric cipher systems use the same key at both ends of the link, source-destination. As such, the enciphering/deciphering algorithm and code is relatively short and the execution time is short. Although this reduces the implementation cost, the key needs to be transported securely. In view of the increased sophistication of attackers and the number of attacks, algorithms have been designed that do not require the same key at both ends of the channel; they required two different keys (one at each end) which are mathematically interrelated. Thus, what is to be known is the abstract mathematical relationship and perhaps some parameters, even if some of them are publicly known. Systems based on this are known as *asymmetric cipher systems*, or *asymmetric cryptography*.

In 1967, it was proposed [13] that it is possible to have an encryption algorithm so that the deciphering key is still difficult to be found by an unauthorized entity even if part of the key is in the public domain.

As intriguing and provocative as this proposal was, in practice it used two keys, a public and a private, which were mathematically interrelated. The first was used to produce the cipher text known as the *public key*. The second, known to the rightful receiver only, was used to decipher the cipher text and it is known as the *private key*.

Because two different keys are used at each side of the channel, cryptographic methods with two different keys at each side of the channel fall in the *asymmetric cryptography* classification or *public key cryptography* because one of the keys is public.

In asymmetric cryptography, security depends on the long time intractability of mathematical manipulations that are required to define the cryptographic keys. However, long time intractability is relative and subjective because what was believed to be extremely (computationally) fast few years ago, is now considered very slow, and so on. For example, if a few years ago a processor executed one instruction in one μsec, it took approximately one year to execute 30 trillion instructions code. However, a modern fast processor at 0.3 nsec per instruction executes the same code in less than three hours. That is, if asymmetric cryptography, besides being complex, is also versatile and flexible to change parameters in a timely manner (that is parameters have a time to live) then cryptography stays ahead of code-breakers.

Simplifying the description of complex asymmetric cryptographic methods requires first defining certain mathematical concepts.

3.3.1 Factorization Problems

The *integer factorization problem* poses the following challenge:

> *given a positive integer n, find its prime factorization.*

That is, find the pair-wise distinct primes P_i such that $n = p_1^{e_1} p_2^{e_2} p_3^{e_3} \ldots p_k^{e_k}$, where the exponents are $e_i \geq 1$. The integer should also be tested for primality or if it is composite; although this test is easier than the factorization, it is as easy for extremely large numbers.

Similarly, the *polynomial factorization problem* is stated as:

Given a polynomial $f(x)$ in $F_q|x|$ with $q = p^m$, find its factorization $f(x) = f_1^{e_1} f_2^{e_2} f_3^{e_3} \ldots f_k^{e_k}$, where $e_i \geq 1$ and $f_i(x)$ is an irreducible polynomial in $F_q|x|$.

The *non-trivial factorization* problem $n = ab$, where the *non-trivial factors a and b* are $1 < a < n$ and $1 < b < n$ and they are not necessarily prime. This problem is solved by splitting n algorithmically to two factors a and b and then test a and b for primality [14].

3.3.2   The Race for Unbreakable Codes

As a consequence of the increased attack activity on data and network security, more advanced cryptosystems and public key protocols have been developed in an effort to strengthen data privacy and protect them from electronic copying, cloning, destroying, and to also deter unauthorized entry into the network [15–17]. The public key cryptography (W. Diffie and M. Hellman, 1976) [18] considers a pair of keys, e and d, where e is *a* publicly available and d is a private key.

Another well known public key cryptosystem is based on elliptic curves and thus is known as the *Elliptic Curve Cryptography* (ECC), which is also gaining popularity and is expected to become a standard for U.S. government data communications. The U.S. National Security Agency (NSA) has already recommended a set of advanced cryptography algorithms known as Suite B for securing U.S. government sensitive and unclassified communications.

The public key protocols in Suite B are

- The Elliptic Curve Menezes-Qu-Vanstone (ECMQV) and the Elliptic Curve Diffie-Hellman (ECDH) for key agreement, and Elliptic Curve Digital Signature Algorithm (ECDSA) for authentication.

- The Advanced Encryption Standard (AES) for data encryption, and

- The Secure Hash Algorithm (SHA) for hashing [19, 20].

In cryptography, a *hash function* takes an arbitrary amount of input and produces an output of smaller and fixed size, known as *digest*. It is impossible to determine the input given the output if the hash function is not known. In addition, the input-output relationship is unique; the output is always the same for the same input.

There are two primary Hash functions, the message digest 5 (MD5) and the secure hash algorithm 1 (SHA1). MD5 was developed by RSA labs and SHA1 by the national security agency (NSA). SHA1 supports messages up to 2^{64} bits at the input and it produces a 160-bit digest.

All Suite B algorithms are consistent with the National Institute of Standards and Technology (NIST) publications. Yet, vulnerabilities or insecurities are not absent from these algorithms. In 2005, it was announced that the SHA-1 algorithm had been broken by a Chinese research team (March 1. 2005, *The Australian*). In fact, *steganalysis* analyzes the hardness of cryptographic codes by taking advantage of weaknesses and by trying to break them using sophisticated methods (including fiddly methods and social engineering). With advanced filtering techniques, it examines encrypted methods and retrieves hidden messages.

In the head-to-head race between cryptography and steganalysis, cryptographers are in search of the "holy grail" of cryptography and use novel and complex algorithms including those with origins in quantum mechanics; conversely, educated codebreakers use sophisticated methods, distributed grid computers and supercomputers, and, when available, quantum computers. That is, it is doubtful that the race will end. In the meantime, cryptography, and the technology that goes into it, will continue to evolve.

3.4 ELLIPTIC CURVE CRYPTOSYSTEMS

The RSA and Diffie-Hellman (DH) public key cryptographic systems are based on number theory and on computational difficulty.

- The RSA algorithm is based on the difficulty of factoring the product of two large primes. A user selects two large prime numbers, and their product becomes the public key. The secrecy of the method depends on the difficulty of factoring a very large number in an attempt to find from the private key from the public key.

- The DH algorithm is based on the difficulty of solving *discrete logarithms* for large finite groups. That is, in a large finite group the solution for x of the equation $a^x = c$, where a and c are publically known, involves difficult discrete logarithms, which they require unrealistic time or they are unsolvable.

In cryptography, the discrete logarithm problem is distinguished in two cases:

- One case assumes that the finite field has a large prime (that is an odd characteristic), and
- The other case assumes the binary or characteristic of 2 (that is, 2^m), which is most useful in binary data.

Although both schemes are formulated differently, they exhibit comparable security as both are based on the difficulty of solving a problem. Clearly, this difficulty may be an insignificant argument with supercomputers or with grid computers, which are expected to compute at petaflops. The *elliptic curve cryptography* (ECC) is another method that capitalizes on the difficulty of calculating the keys, yet more difficult to the un-initiated to elliptic curves crypto-attacker.

The year 1985 was very productive in elliptic curve cryptography: Hendrik Lenstra proposed the elliptic curve factorization whereas N. Koblitz [21] and V. Miller [22] proposed independently a general cryptographic method called *Elliptic Curve Cryptography* that was primarily designed to support digital signature algorithms, to exchange encryption keys, and to authenticate content and integrity while reducing the cipher key length and performing operations faster. Moreover, in 1991 the Menezes-Okamoto-Vanstone sub-exponential time algorithm for super-singular curves was proposed [23] as well as many others notable papers that added ECC in the cryptographers toolbox [24–33]. Since then, ECC was elaborated by the ANSI X9F1 working group, published in the IEEE 1363 standard and the SEC 1 standard and it evolved to a Suite B of public-key algorithms in the Cryptographic Message Syntax (CMS) and; the processing of the ECC algorithm is based on ANSI X9.62 [34, 35].

CMS is an independent cryptographic algorithm, yet its content type "*SignedData*" supports the creation of the ECC-based digital signature algorithm (ECDSA). Similarly, its content type "*EnvelopedData*" supports ECC-based public encryption key agreements for content encryption, and its content type "*AuthenticatedData*" generates pair-wise encryption keys for content authentication and integrity.

3.4.1 A Primer on Elliptic Curves and Elliptic Curve Factorization

In search of more complex yet more efficient asymmetric cryptographic methods, elliptic curves have been introduced. These elliptic mathematical expressions are of the general form $E^{a,b}(GF(n))$: $y^2 = f(x)$ or $F(x, y) = y^2 - f(x)$ (where G denotes Group and F denotes Field), and they are *smooth* at any point of the curve; that is, the slope at any point on the curve or the first differential of F with respect to x and to y does not vanish. One may envision the relationship $F(x, y)$ as projective curves on the plane through the origin in three-dimensional space of an equivalent class $(x:y:z)$ of curves, such that they are reduced to regular (x, y) coordinates. Then, the set of points $(x:y:0)$ is called the *line at infinity*. As an example, consider a cone. Then, all conics can be projected on a plane through the origin and be transformed into each other by a linear transformation. Similarly, an elliptic curve is a two-variable plane curve, on which a group of points in the affine plane satisfies a certain cubic equation (Figure 3.7). If the field is in the complex space, the elliptic curve becomes a two dimensional surface (that is, it is visualized in three dimensions). For example, a torus is a two dimensional surface in the complex space, and the addition of pairs of points in the complex space becomes the addition of pairs of angles.

In elliptic curve cryptography, the general form $y^2 + axy + by = x^3 + cx^2 + dx + e$ has been deployed (in the literature, the notation $y^2 + a_1xy + a_3y = x^3 + a_2x^2 \; a_4x + a_6$ is also encountered). The points x and y that satisfy the elliptic curve, together with a point at infinity 0 (see above) that satisfy the addition law, and thus, mathematically speaking, these points form an abelian group whose identity element is 0. Based on this, multiplication of points P on the elliptic curve by a scalar k is a convenient feature because these points are computed using the *addition rule*. It is these special pairs of points on elliptic curves that can be used as asymmetric keys in public-key cryptography (Figure 3.8).

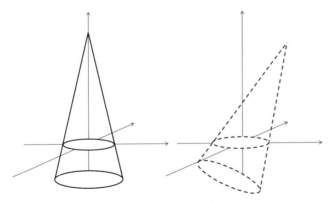

Figure 3.7. Simplified conic through the origin.

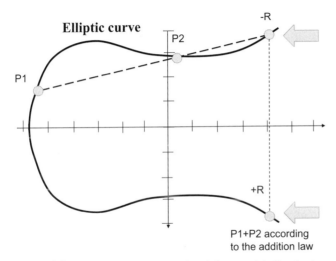

Figure 3.8. An elliptic curve; two points define a straight line that intersects the elliptic curve at point −R (the first key); its mirror image determines the second point (key +R).

In cryptography, elliptic curves are defined with finite fields and are used as the generating function, such as $y^2 = x^3 + ax + b$(modulo-p). Hence, the general form of the elliptic curve E over a field F $y^2 + axy + by = x^3 + cx^2 + dx + e$ is *smooth*, meaning that there is no point on the curve at which both partial derivatives vanish.

Consider two cases of elliptic curve groups. One is over the field of real numbers $F(n)$, and the other over the field $F(p)$ where p is a prime, and over the field $F(2^m)$ which is a binary representation with 2^m elements.

For real numbers, we distinguish two cases:

- If the field F is not a characteristic of 2, then $a = b = 0$ and $y^2 = x^3 + cx^2 + dx + e$.

- If F is a characteristic of 2 (this case is interesting in digital arithmetic), then there are two sub-cases, the *supersingular* and the *non-supersingular*:
  - for a finite field F, if the order of the elliptic curve minus 1 is divisible by a prime number then the elliptic curve is called *supersingular*, in which case $a = 0$ and $y^2 + by = x^3 + cx^2 + dx + e$
  - If it is not divisible by a prime number, then it is called *non-supersingular*, in which case $b = 0$ and $y^2 + axy = x^3 + cx^2 + dx + e$

Most elliptic curves are non-supersingular. However, it is the supersingular elliptic curves that are most useful in cryptography for which it is most difficult to solve the discrete logarithm problem (that is, given two points P and Q, find an integer k such that $P = kQ$).

The elliptic curve factoring algorithm considers a random elliptic curve group over Z_p. The order of such group is roughly uniformly distributed in the interval from $\left(p + 1 - 2\sqrt{p}\right)$ to $\left(p + 1 + 2\sqrt{p}\right)$. If the order of the group is smooth with respect to some pre-selected bound, the elliptic curve algorithm most likely will find a non-trivial factor of n, else it will fail.

The elliptic curve, the elliptic curve domain parameters, and the base point are the necessary information needed to define public key cryptographic schemes that are collectively called "*elliptic curve cryptography*" (ECC).

In the sections that follow we look into elliptic curves and elliptic cryptography in more depth.

3.4.1.1 Elliptic Curves Over Real Numbers

In this case, the defining elliptic curve equation is $E^{a,b}(GF(n))$: $y^2 = x^3 + ax + b$, where x, y, a and b real numbers. If there are no repeated factors in $y^2 = x^3 + ax + b$ (in practice $4a^3 + 27b^2 \neq 0$), then the elliptic curve forms a group. That is, each value of a and b yields a different elliptic curve, which is symmetric about the x axis. Figure 3.9 illustrates an example of three representative elliptic curves of a group.

The following pseudo-code generates a set of elliptic curves $y^2 = x^3 + ax + b$ for different values of a and b.

```
For b = (0 to 4, step 1);
    For a = (0 to 4, step 1);
        For x (−10 to 10, step 0.5);
            If x³ + ax + b ≥ 0; test if y value is on the elliptic curve
            w = x³ + ax + b
            y₁ = +√w, y₂ = −√w; find y
            Plot (x, y₁),(x, y₂); two symmetric points (or a single point if y₁ = y₂)
            Else,
        return
        Print (a, b)
    return
return
```

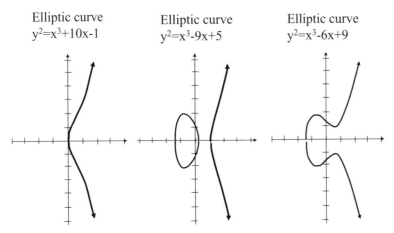

Elliptic curve
y²=x³+10x-1

Elliptic curve
y²=x³-9x+5

Elliptic curve
y²=x³-6x+9

Figure 3.9. Three representative elliptic curves of a group.

An elliptic curve over real numbers consists of all points on the curve together with the special point O, the *point at infinity*, the purpose of which becomes clear if we notice that elliptic curves are symmetric about the axis x. Thus, because elliptic curve groups are additive, if we were to add two points P and P' that are symmetric around the x axis, the result would be "null" or the point O. Although the point at infinity O has no affine coordinates, it is convenient to represent it with a pair of coordinates that do not satisfy the defining equation.

For example, $O = (0, 0)$ if $b \neq 0$ and $O = (0, 1)$ otherwise. *Affine coordinates* are the coordinates (x, y) of a locus. For example, the elliptic curve for which $a = -3$ and $b = 3$ has the (x, y) loci on the curve {(−2.1, +/−0.20), (−2.0, +/−1), (−1.0, +/−2.24), (0.0, +/−1.73), (1.0, +/−1), (2.0, +/−2.24), (3.0, +/−4.58), (4.0, +/−7), (5.0, +/−10.63)}.

The additive property of elliptic curves is examined in the remainder of this section.

Similarly, some of the loci on the elliptic curve with $a = -6$ and $b = 9$, Figure 3.10, rounded to the second decimal point are: {(−2.93, ±1.26), (−0.80, ±3.65), (0.93, ±2.06), (2.93, ±4.06), (5, ±10.2)}.

Consider two points P and Q on the elliptic curve of Figure 3.10. Point P has coordinates (x_1, y_1) and its symmetric point on the curve and about the axis x is $-P = (x_1, -y_1)$. Point Q has coordinates (x_2, y_2) and it is such that $Q \neq -P$. Adding the two points P and Q geometrically is equivalent to drawing a straight line through points P and Q until it intersects the elliptic curve at a third point, R'. Then, the symmetric of point R' around the axis x, point R, is the solution to $P + Q = R = (x_3, y_3)$ (Figure 3.11).

Algebraically, the point $R = P + Q$ on the curve with coordinates (x_3, y_3) for the elliptiic curve $y^2 = x^3 + ax + b$ is found as follows [36]:

Elliptic curve
y²=x³-6x+9

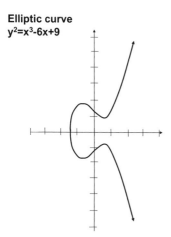

Figure 3.10. Elliptic curve $y^2 = x^3 - 6x + 9$.

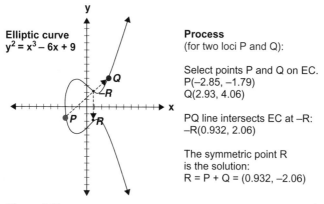

Elliptic curve
$y^2 = x^3 - 6x + 9$

Process
(for two loci P and Q):

Select points P and Q on EC.
P(−2.85, −1.79)
Q(2.93, 4.06)

PQ line intersects EC at −R:
−R(0.932, 2.06)

The symmetric point R
is the solution:
R = P + Q = (0.932, −2.06)

Figure 3.11. The process for adding two points on the elliptic curve $y^2 = x^3 - 6x + 9$.

$$x_3 = S^3 - x_1 - x_2, \text{ and}$$

$$y_3 = S(x_1 - x_3) - y_1,$$

$$S = (y_2 - y_1)/(x_2 - x_1) \text{ if } P \neq Q, \text{ or } S = (3x_1^2 + a))/2y_1$$
$$\text{if } P = Q \text{ (that is, } P \text{ is a double point)}$$

where S is the slope of the line through the points P and Q (when $P \neq Q$) or the slope of the tangent at point P ((when $P = Q$).

Similarly, the coordinates (x_3, y_3) of $P + Q = R$ for the elliptiic curve $y^2 + xy = x^3 + ax^2 + b$ are defined as follows:

For $P \neq Q$:

$$x_3 = [(y_1 + y_2)/(x_1 + x_2)]^2 + (y_1 + y_2)/(x_1 + x_2) + x_1 + x_2 + a, \text{ and}$$
$$y_3 = [(y_1 + y_2)/(x_1 + x_2)](x_1 + x_3) + x_3 + y_1, \text{ and}$$

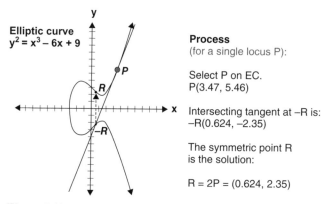

Figure 3.12. The process of finding the pointe −R and +R at a single point.

For $P = Q$ (that is, P is a double point):

$$x_3 = x_1^2 + b/x_1^2, \text{ and}$$
$$y_3 = x_1^2 + (x_1 + y_1/x_1)x_3 + x_3.$$

EXAMPLE 3.3: *Consider the elliptic curve $y^2 = x^3 - 6x + 9$, Figure 3.11. In this case, point $P = (-2.85, -1.74)$, $Q = (2.93, 4.06)$ and $P + Q = R = (0.932, -2.06)$.*

Now, if a single point P on the elliptic curve is selected then the tangent at P intersects the elliptic curve at another point −R from which the symmetric point R is the solution to $2P = R$ (that is, point P is a *double point*), Figure 3.12.

3.4.1.2 Elliptic Curves Over Real and Prime Numbers

Working with real numbers produces imprecise results due to the rounding of numbers which cause errors. For instance, $5/3 = 1.6666666666*$ which one may write 1.667 or 1.6667 if accuracy to the third or fourth decimal is sufficient; this introduces an error of approximately 0.0003. Similarly, the product 1.35×2.97 produces 4.0095, which if rounded to the second decimal is 4.01 introducing an error of 0.0005. In cryptography, precision is very critical and therefore fields with integers, primes and powers of 2, are more convenient. The case which produces very large output changes to minute input parameter changes (changes on the 4th and beyond this decimal place) is the subject of chaotic processes, on which chaos *cryptography* is based and is examined in Chapter 5.

In the prime case, the defining equation has the form $E^{a,b}(GF(p))$: $y^2 = x^3 + ax + b$, where a and b are constants in $F(p)$ such that $4a^3 + 27b^2 \bmod p \neq 0$, where p is a prime number in the range 0 to $p - 1$, and the right side of the equation has no repeated factors; $E^{a,b}(GF(p))$ denotes the group of curves over the field $F(p)$ for the prime case. Thus, an elliptic curve consists of all points (x, y) that satisfy the elliptic curve equation modulo p, $y^2 \bmod p = (x^3 + ax + b) \bmod p$, along with a special point at infinity O; in addition, the relationship $4a^3 + 27b^2 \bmod p \neq 0$ must be satisfied. An elliptic curve over prime consists of the points on the curve together with the point O. The negative of a point $P = (x_P,\ y_P)$ on the elliptic curve is point $-P = (x_P,\ -y_P \bmod p)$.

As in the case with real numbers, the addition of two points P and Q on the elliptic curve $E^{a,b}(GF(p))$ produces a third point R with coordinates $R = (x_R,\ y_R)$:

$$x_R = S^2 - x_P - x_Q \bmod p, \text{ and } y_R = -y_P + S(x_P - x_R) \bmod p,$$

where $S = (y_R - y_Q) / (x_P - x_Q) \bmod p$ is the slope of the straight line that connects points P and Q.

If there is only one point P on the curve at which the tangent is considered (also alled double point), then the point R is calculated from $2P = R$, in which case $(x_R,\ y_R)$ are:

$$x_R = S^2 - 2x_P \bmod p, \text{ and } y_R = -y_P + S(x_P - x_R) \bmod p,$$

where $S = (3x_P^2 + a) / (2y_P) \bmod p$ is the slope of the tangent at point P.

Two popular elliptic curves defined in ANSI, SECG, FIPS standards, and in the prime case $Ea,b(GF(p)):y^2 = x^3 + ax + b$, and in the binary case $Ea,b(GF(2^m)):y^2 + xy = x^3 + ax^2 + b$.

EXAMPLE 3.4: *Consider a simple elliptic curve in F(23) with a = 1 and b = 0, $y^2 = x^3 + x$. Examine if point (x, y) = (9, 18) is a point on the elliptic curve.*

First, the coefficient relationship $4a^3 + 27b^2 \bmod p$ is examined: since $4 \bmod 23 \neq 0$ this condition is satisfied.

Now, to examine if the point $(x,\ y) = (9, 18)$ is a point on the elliptic curve substite x and y by the coordinate values:

$18^2 \bmod 23 = (9^3 + 9) \bmod 23$ or $324 \bmod 23 = 738 \bmod 23$ or $2 = 2$; this confirms that point $(9, 18)$ is a point on the selected elliptic curve.

Exercise: There is a total of 23 points that satisfy the elliptic curve defined in the aforementioned example. Using the previous example, verify that points $(1, 5)$, $(9, 5)$, $(11, 13)$, $(17, 13)$, and $(19, 1)$ are among them.

3.4.1.3 Elliptic Curves Over Binary Numbers

In the binary case *field size* is $q = 2^m$ and the defining equation has the form $E^{a,b}(GF(2^m)): y^2 + xy = x^3 + ax^2 + b$, where a and b are unequal constants in $F(2^m)$, $b \neq 0$, and m is a large integer. In this notation, $E^{a,b}(GF(2^m))$ denotes the group of curves over the field $F(2^m)$ for the binary case. As in previous cases, $F(2^m)$ consists of all points on the elliptic curve and the point at infinity O. Notice that points in the binary case are expressed in their binary representation with the two symbols (1,0) where m denotes the number of symbols in the string. For example, if $m = 4$ then 0110 is one of the sixteen (2^4) possible strings. In addition, since we deal with binary strings, calculations are executed with the modulo-2 or Exclusive Or (XOR) function. In a realistic cryptographic system, m is very large, such as >100 in order to be meaningful; NIST defines a field $GF(2^{521} - 1)$; that is, $n = 521$.

In the binary case, the form $E^{a,b}(GF(2^m))$ takes a more general expression where the coefficients a and b are binary strings and thus they are replaced by the binary form g^m (g is known as the *generator*) as:

$$E^{a,b}\left(GF\left(2^m\right)\right): y^2 + xy = x^3 + g^m x^2 + g^n.$$

Then, for the simple elliptic curve $E^{a,b}(GF(2^2))$ with generator $g = (0010)$, the powers of g are in binary form:

$$g^0 = (0001),\, g^1 = (0010),\, g^2 = (0100),\, g^3 = (1000),$$
$$g^4 = (0011),\, g^5 = (0110),\, g^6 = (1100),\, g^7 = (1011),$$
$$g^8 = (0101),\, g^9 = (1010),\, g^{10} = (0111),\, g^{11} = (1110),$$
$$g^{12} = (1111),\, g^{13} = (1101),\, g^{14} = (1001),\, g^{15} = (0001);\text{ that is } g^0 = g^{15}$$

EXAMPLE 3.5: *Consider the elliptic curve $E^{a,b}(GF(2^2)): y^2 + xy = x^3 + g^4 x^2 + g^0$ ($a = g^4$, and $b = g^0$) and the generator $g = (0010)$. Verify that the point $(x, y) = (g^3, g^8)$ lies on the elliptic curve.*

Replacing for x and y, there is:

$$\left(g^8\right)^2 + g^3 g^8 = \left(g^3\right)^3 + g^4 \left(g^3\right)^2 + g^0,\text{ or}$$

$g^{16} + g^{11} = g^9 + g^{10} + g^0$, or and replacing with the corresponding binary codes (notice that when the exponent becomes greater than 15, they are modulo reduced and thus $g^{16 \bmod 15} = g^1$):

$(0010) + (1110) = (1010) + (0111) + (0001)$, and performing XOR operations between them, then

$$(1100) = (1100).$$

In the binary case of elliptic curves, there is a negative point R so that $R + (-R) = O$ and $P + O = P$, where O is the point at infinity. Moreover, two points P and Q are added to produce a point R, $P + Q = R$, the coordinates of which are (remember that all calculations are modulo-based):

$$x_R = S^2 + S + x_P + x_Q + a,\text{ and } y_R = S(x_P + x_R) + x_R + x_P,$$

where $S = (y_P - y_Q) / (x_P + x_Q)$.

Similarly, if the point P is double at which the tangent is drawn, the coordinates of the point R, $2P = R$, is the calculated from:

$$x_R = S^2 + S + a,\text{ and } y_R = x_P^2 + (S+1)x_R,$$

where $S = x_P + y_P / x_P$

3.4.1.4 Cyclic Groups

A useful concept that helps to understand elliptic curve manipulation in the context of cryptography is the *cyclic groups*. Raising an element a to the power n (where n is a finite positive or negative integer, or 0): $a^0 = 1$, $a^1 = a$, $a^2 = aa$, and $a^n = a^k a^m = a^{k+m}$, $a^n = a^{n-1}a$, and $a^n = aaa \dots a$ (n times); the latter in additive notation is written as nxa or $n.a$.

Consider the group of elements a^0, a^1, a^2, a^3, ... , a^k, ... , a^{n-1}; all elements in the group $G(a^k)$ are powers of a, where k is a finite integer, $k = 0$ to $n - 1$. In addition, consider that the modulus n of the last term of the group, a^{n-1}, results to the identity element or simply the unit; that is,

$$a^{n-1} \bmod n = 1,$$

which in simplified notation (for which mod$n$ is implicit) is

$$a^{n-1} = 1.$$

When a^{n-1}mod$n = 1$, the group is termed *cyclic group of order n* or the field size is n, F_n; that is, the first and the last element in a cyclic group is the unit.

In this case, a is said to be a *generator* of the cyclic group of size n, denoted C_n, or that the group is generated by a.

For example, the field F_{23} has a size of 23 elements from 0 to 22 (notice that 23 is a prime). The prime 23 acts like a field, F_{23}, and the 23 elements of the group for the generator $a = 5$ (assuming mod23 operation for each term) are: $5^0 = 1$, $5^1 = 5$, $5^2 = 2$, $5^3 = 10$, $5^4 = 4$, $5^5 = 20$, $5^6 = 8$, $5^7 = 17$, $5^8 = 16$, ... , $5^{21} = 14$, and $5^{22} = 1$; that is, the field F_{23} returns to 1 after 23 terms. Cyclic groups are abelian, but all abelian groups are not cyclic.

3.4.2 Applicability of Elliptic Curves to Cryptography

The applicability of elliptic curves to cryptography using simplified mathematical language is illustrated if we consider the algebraic closed elliptic curve and a straight line that cuts the elliptic curve at three points, as already shown. If the line is tangent to the elliptic curve, the tangent point is considered twice. Then, if two of the intersection points are rational, so is the third, and if two of the points are known, it is possible to compute the third. All solutions of the equation together with a point at infinity form an *abelian group* with the point at infinity as the identity element. An abelian group G is said to be a *commutative group* if $a + b = b + a$ and $a * b = b * a$ for all (a, b) in G; this means that it does not matter in which order a binary operation is performed. Thus, if the coordinates x and y are chosen from a large finite field, the solutions also form a finite abelian group. In common language, the third intercepting point is calculated from the two previous ones using the additive operation $P + Q = R$, as already described.

The ECC method suggested by Koblitz and by Miller was based on the difficulty of solving the elliptic curve discrete logarthm problem (ECDLP), which is stated as:

> Given a prime power p, where F_p denotes the finite field containing p elements, and an elliptic curve, E, defined over a finite field F_p determine the integer n in the range (0, n − 1), such that a point P on the elliptic curve $E(F_p)$, and another known point Q on the elliptic curve $E(F_p)$, are related by nP = Q; that is, point Q is an integer multiple of point P provided that such integer exists.

It is believed that solving the ECDLP (that is, if P and n are known then Q can be generated) is computationally more difficult than solving the discrete logarithm problem in the Diffie-Hellman (DH) method, although algorithms to solve ECDLP have been proposed. However, for comparable or better security, the secret cryptographic key derived from elliptic curves is much shorter than the DH method.

EXAMPLE 3.6:   *Consider the elliptic curve $y^2 = x^3 + 9x + 17$ over F_{23}. What is the discrete logarithm k of Q = (4, 5) to the base P = (16, 5)?*

Using the brute force approach, one would calculate all possible multiples of P starting with k = 1 and progressing with k = 2, and so on until the solution that satisfies kP = Q = (4, 5) is found. Doing so, the first few multiples are: P = (16, 5), 2P = (20, 20), 3P = (14, 14), 4P = (19, 20), 5P = (13, 10), ... , 9P = (4, 5). Thus, the discrete logarithm of Q to the base P is k = 9.

The number of points or loci on elliptic curves that are useful in elliptic cryptography is of importance as also is the number of points that make the elliptic curve discrete logarithm breakable, so that they are avoided. In general, the elliptic curve may be vulnerable to attacks if the number of points on E over F is the same as the number of elements of F.

Suppose that there is a^m such that m is written as m = nQ + R, where $0 \le R < n$ for some integer Q; remember that m = nQ + R is also expressed as R = mmodn. Replacing m with the latter, one obtains $a^m = a^{nQ+R} = a^{nQ}a^R = (a^n)^Q a^R = (1)^Q a^R = a^R$. This means that all powers are in the range [0 to n], in which case the cyclic group of order n is written $C_n = <a>$, where a is the generator and n is a prime. Thus, the additive cyclic group is considered similar to a multiplicative group of powers of an integer amod(prime n), and the problem of powers of finding k given points (kC and C) constitutes the *elliptic curve discrete logarithmic problem* (ECDLP); if prime numbers are used, it is known as the *prime case*, and if binary numbers (2^n), as the *binary case*.

3.4.2.1 ECC Procedures

To set up an elliptic curve cryptosystem, one needs to first establish the system parameters. This part consists of the steps:

- Select a finite field and a representation of the elements in it.
- Select an elliptic curve and the *generator* point G on the curve.
- Generate the public-private key pairs. These keys consist of a random, secret integer k, which acts as the private key, and the multiple of the generator point, kG, that acts as the public key.

It is the number of points N in the group of elliptic curve cryptosystems that renders difficult to compute the private key k from the public key kG, thus offering higher security. This part requires a sequence of elliptic curve additions where each addition consists of several arithmetic operations in the finite field.

3.4.2.2 Comparing ECC with DH and with RSA

In contrast to ECC, the RSA cryptosystem requires:

- no system parameters, but two primes of appropriate size, and
- the generation of the public-private key pairs, the public modulus n from the product of the two primes, a process which is less computationally intensive than setting up the elliptic curve system parameters [37].
- Additionally, the secret exponent d is computed, although this computation is insignificant when compared to generating the primes.

In RSA cryptosystems, security lies on the difficulty to compute the length of the modulus and a sequence of modular multiplications that offers security.

To summarize, the elliptic curve cryptography (ECC) is considered to be more computationally difficult, offering higher security than its predecessors with equal or shorter keys. When comparing the level of security offered, an ECC over a 160-bit field $GF(2^{160})$ offers roughly the same security as a 1024-bit RSA modulus2, and a $GF(2^{136})$ as a 768-bit RSA. Comparing key sizes recommended by NIST [38, 39] for the Diffie-Hellman, RSA and ECC, there is a dramatic difference in key length: 1024 vs 163; 3072 vs 256; 7680 vs 384; and 15,360 vs 512, respectively.

Does this exhaust the subject of cyclic groups? Or state all properties of cyclic groups? The answer is NO to both questions.

The objective of this section was not to provide a complete mathematical treatment in few pages (this is the subject of math books) but to identify the complexity of ECC and ECDLP and trigger further interest.

Current active research in this field is the minimization of key size, elliptic curves without hard points (that result in keys) that can be attacked, and also to identify those hard points on elliptic curves that can be attacked to avoid them sooner in ECC. The hardest ECC broken to date has a key that provides about 55 bits of security. For the prime field case, ECC was broken in 2003 using 10,000 Pentium class PCs running continuously for over 540 days, and for the binary field case it was broken in 2004 using 2600 computers (on a grid) for about 510 days. In the same year, NSA licensed Certicom's patents on EC cryptography for $25 million, and it also announced the Suite B as the US government standard, which uses Elliptic Curve Diffie-Hellman (ECDH) and Elliptic Curve Menezes-Qu-Vanstone (ECMQV) for key agreement, and uses Elliptic Curve Digital Signature Algorithm (ECDSA) for signature generation/verification.

3.5 THE RSA ALGORITHM

This algorithm, developed by Rivest, Shamir and Adleman (hence RSA algorithm) is widely used in many cryptographic applications; it is based on prime numbers and on Euler's and Fermat's theorems (see Chapter 2).

According to RSA algorithm [40, 41], each entity chooses two prime numbers, p and q, and calculates the products $N = pq$ and $\Phi = (p - 1)(q - 1)$. Then, it selects a number k such that $gcd(k, \Phi) = 1$. Then, it finds a unique integer k' such that $kk' - 1$ is evenly divisible by $(p - 1)(q - 1)$; that is, $kk' = 1 \ mod(\Phi)$.

This algorithm considers the key k public with which the text is ciphered and the k' private. The cipher text then is $C = M^k modn$ and the deciphered original text is $M = C^k modn$.

EXAMPLE 3.7

- Assign a number to each letter of the alphabet, as

A	B	C	D	E	F	G	H	I	J	K	L	M	N	O	P	Q	R	S	T	U	V	W	X	Y	Z
1	2	3	4	5	6	7	8	9	10	11	12	13	14	15	16	17	18	19	20	21	22	23	24	25	26

- Then, divide the message to be encrypted into groups of letters; the size of group is chosen so that two groups are not identical. For example, assume that a message consists of the word ABCD. In this case, the length of the group is a single letter such as A, B, C, and D. Then, the letters are replaced by their corresponding numbers in the alphabet, as A B C D => 1 2 3 4.
- The algorithm continues with choosing a number N which is the product of two prime numbers, p and q. In this case, the selections are $p = 3$ and $q = 7$, and thus $N = 3 \times 7 = 21$, and $\Phi = 2 \times 6 = 12$.
- Then, a prime number k is selected that is within the product $(p - 1)(q - 1)$. In this example the selection is $k = 7$ which is within 1 and 12. The prime number $k = 7$ constitutes the cipher key.
- Then, the message is encrypted as follows: each number is raised to the power of the key times (mod n): $1^7 mod15$, $2^7 mod15$, $3^7 mod15$, $4^7 mod15$, or $1 mod15$, $128 mod15$, $2187 mod15$ and $16384 mod15$, or 14, 8, 12, 4, and thus the encrypted message is N H L D

To decrypt this message:

- The recipient of the encrypted message N H L D converts it to letter numbers of the alphabet, 14, 8, 12, and 4.
- Then, a process starts to calculate the decipher key. To do so, the recipient has to find a value k' such that $kxk' - 1$ is evenly divisible by $(p - 1) \times (q - 1)$,

or $[kxk' - 1] / [(p - 1) \times (q - 1)] = P$ where P is an even number. In modular notation this is equivalent to $kxk' - 1 = 0 mod[(p - 1) \times (q - 1)]$.

- Then, k' constitutes the decipher key.

In our trivial example, $(p - 1) \times (q - 1) = 2 \times 6 = 12$ and $(7xk' - 1) / 12 = P$ yields $k' = 7$ as $P = 4$; that is, the cipher key and the decipher key turn out to be the same. Unfortunately, this is because in this illustrative example we used very small prime numbers. However, if one uses large prime numbers and a computer to do the arithmetic, then the two keys k and k' are not the same and thus the method is asymmetric (as an exercise, use the numbers $k = 11$, $p = 13$ and $q = 19$ and find k', or use any large numbers from the table of prime numbers in Appendix A). The actual RSA algorithm involves prime numbers p and q that have the same bit-length and are very large so that so that the computation $n = pq$ is infeasible; for example, p and q should be 512 bits long if a 1024-bit modn is to be used.

- Now, knowing the key k', the decrypted message is recovered by following a similar process as in the process that encrypted it. Then, the corresponding numbers of the cipher text are raised to the power of $k'modn$ and the result is the deciphered text:

Based on the last step, our example continues as: 14^7mod15, 8^7mod15, 12^7mod15, 4^7mod15, or $105413504mod15$, $2097152mod15$, $35831808mod15$, and $16384mod15$, which yield 1, 2, 3, and 4, and they produce the deciphered and original text A B C D.

In conclusion, the RSA algorithm is based on difficult mathematical computability to strengthen the privacy of a message and to also make the job of a bad actor difficult. However, all an eavesdropper needs to know is the cipher k and N, which is the product of two prime numbers, p and q, from which it should be difficult but not impossible to compute with a supercomputer. Therefore, if the sender and the receiver can use fast computers for ciphering and deciphering, so can a sophisticated eavesdropper. The latter is not underestimated and in fact, RSA had an open challenge at $100 to whoever could break its code RSA-129, an algorithm using a 129-bit key. In the summer of 1994, an international team, dubbed the *wisecrackers*, was able to crack the code using 1,600 computers that were distributed in every continent and worked for approximately one year.

The asymmetric-key RSA algorithm as explained is substantially slower that symmetric key algorithms. This has been improved by using special algorithms and also customized hardware. Thus, RSA in practice has become a popular algorithm, has been patented, and has been incorporated in standards for several security applications.

3.6 KEY MANAGEMENT

As already elaborated, the secrecy of information is as good as the secrecy (or protection against unauthorized disclosure) of the cryptographic keys and also as good as its unbreakability (or strength). *Key management* sets the rules and practices for the management of cryptographic keying material; that is, key protection against modification, key storage, key distribution, key destruction, security planning requirements, usage of cryptographic features of cryptographic systems, and user adherence to these rules [42–45].

Some of the issues (most of which have already been elaborated upon) that key management addresses:

- Which are the security services and the key types in cryptographic mechanisms?
- How are keys classified by function?
- In which state keys are during their lifetime?
- What is the key usage?
- What is the crypto-period length?
- How is key management supported within an organization?
- How can key management address the security issues of already deployed and in use cryptographic systems?
- And many more issues, such as key validation, accountability, audit, key size selection, and so on.

REFERENCES

1. J. DAEMEN, R. GOVAERTS, and J. VANDEWALLE, "A New Approach to Block Cipher Design," *Fast Software Encryption, Cambridge Security Workshop Proceedings*, Volume 809 of Lecture Notes in Computer Science (Ross Anderson, ed.), pp. 18–32. Springer-Verlag, 1994.
2. US Department of Commerce/NIST, FIPS PUB 46-3, Data Encryption Standards (DES), Oct 25, 1999.
3. US Department of Commerce/NIST, FIPS PUB 81, DES Modes of Operation.
4. J. KELSEY, B. SCHNEIER, and D. WAGNER, "Related-Key Cryptanalysis of 3-WAY, Biham-DES, CAST, DES-X, NewDES, RC2, and TEA", *ICICS '97 Proceedings*, Springer-Verlag, November 1997.
5. J. KELSEY, B. SCHNEIER, and D. WAGNER, "Key-Schedule Cryptanalysis of 3-WAY, IDEA, G-DES, RC4, SAFER, and Triple-DES", http://www.counterpane.com/key_schedule.html
6. J. DAEMEN and V. RIJMEN, *AES Proposal: Rijndael*, AES Algorithm submission, September 3, 1999, available at http://www.nist.gov/CryptoToolkit
7. J. NECHVATAL, et al., Report on the Advanced Encryption Standard (AES), NIST, October 2, 2000, available at http://www.nist.gov/CryptoToolkit
8. US Department of Commerce/NIST, FIPS PUB 197, *Advanced Encryption Standard*, November 26, 2001.
9. http://csrc.nist.gov/publications/fips/fips197/fips-197.pdf
10. http://csrc.nist.gov/encryption/aes//rijndeal/Rijndael.pdf
11. R.L. RIVEST, *The RC5 Algorithm*, MIT Laboratory for Computer Sciences, rivest@theory.lcs.mit.edu.
12. M. GARDNER, *The Colossal Book of Mathematics: Classic Puzzles, Paradoxes, and Problems*, Chapter 9, *"The Helix"*, pp. 117–127, W. W. Norton, New York, 2001.
13. W. DIFFIE and M.E. HELLMAN, "New Directions in Cryptography", *IEEE Transactions on Information Theory*, vol. 13, pp. 644–654, November, 1967.

14. A. MENEZES, P. van OORSCHOT, and S. VANSTONE, *Handbook of Applied Cryptography*, CRC Press, 1996.
15. Anti-Phishing Act of 2005, "*A bill to criminalize Internet scams involving fraudulently obtaining personal information, commonly known as phishing*", March 1, 2005; http://www.govtrack.us/congress/billtext.xpd?bill=h109-1099.
16. ELIZABETH II, UK, Fraud Act 2006, Chapter 35, 8th November, 2006.
17. S.V. KARTALOPOULOS, "Optical Network Security: Countermeasures in view of channel attacks", Unclassified Proceedings of Milcom 2006, October 23–25, 2006, Washington, D.C., on CD-ROM, ISBN 1-4244-0618-8, Library of Congress 2006931712, paper no. US-T-G-404.
18. W. DIFFIE and M.E. HELLMAN, "New Directions in Cryptography", *IEEE Transactions on Information Theory*, vol. 13, Nov 1967, 644–654.
19. FIPS 180-2, *Secure Hash Standard (SHS)*, August 2002.
20. FIPS Pub 198, The Keyed-Hash Message Authentication Code (HMAC), March 2002.
21. N. KOBLITZ, *Elliptic curves cryptosystems*. Mathematics of Computation, vol. 48, 1987, 203–209.
22. V.S. MILLER, "Uses of elliptic curves in cryptography", Advances in Cryptology CRYPTO'85, Lecture Notes in Computer Science, vol. 218, Springer-Verlag, pp. 417–426, 1986.
23. A. MENEZES, T. OKAMOTO, and S.A. VANSTONE, "Reducing elliptic curve logarithms to logarithms in a finite field", Proceedings of the 23rd Annual ACM Symposium on Theory of Computing, pp. 80–89, 1991.
24. N. KOBLITZ, *Introduction to Elliptic Curves and Modular Forms*. Graduate Texts in Math. No. 97, 2nd ed., Springer-Verlag, New York, 1993.
25. V.S. MILLER, "Use of elliptic curve in cryptography", *Advances in Cryptology*, Crypto'85, pp. 417–426, Springer-Verlag, 1986.
26. N. KOBLITZ, *Algebraic Aspects of Cryptography, Algorithms and Computation in Mathematics*, Vol. 3, Springer-Verlag, New York, 1998.
27. D. HANKERSON, A. MENEZES, and S.A. VANSTONE, *Guide To Elliptic Curve Cryptography*, Springer-Verlag, 2004.
28. I. BLAKE, G. SEROUSSI, and N. SMART, *Elliptic Curves in Cryptography*, London Mathematical Society 265, Cambridge University Press, 1999.
29. I. BLAKE, G. SEROUSSI, and N. SMART, editors, *Advances in Elliptic Curve Cryptography*, London Mathematical Society 317, Cambridge University Press, 2005.
30. L. WASHINGTON, *Elliptic Curves: Number Theory and Cryptography*, Chapman & Hall / CRC, 2003.
31. N. KOBLITZ, A. MENEZES, and S. VANSTONE, "*The state of elliptic curve cryptography*", Design, Codes and Cryptography, vol. 18, pp. 173–193, 2000.
32. A. MENEZES, *Elliptic Curve Public Key Cryptosystems*, Kluwer Academic Publishers, 1993.
33. A. MENEZES, *Elliptic Curve Cryptosystems*, CryptoBytes, Vol.1 No.2, Summer 1995.
34. S. BLAKE-WILSON, D. BROWN, P. LAMBERT, RFC 3278, "*Use of Elliptic Curve Cryptography (ECC) Algorithms in Cryptographic Message Syntax (CMS)*", April 2002.
35. SEC#1, "Elliptic Curve Cryptography", Standards for Efficient Cryptography Group, 2000. Also available from www.secg.org/collateral/sec1.pdf
36. N. KOBLITZ, A. MENEZES, and S. VANSTONE, "The state of elliptic curve cryptography", Design, Codes and Cryptography, vol. 18, 173–193, 2000.
37. N. DEMYTKO, A new elliptic curve based analogue of RSA, *Advances in Cryptology*, Eurocrypt'93, pp. 40–49, Springer-Verlag, 1994.
38. NIST, FIPS Pub 186-2, Digital Signature Standard, January 2000.
39. NIST, FIPS Pub 186-2 change notice, *Digital Signature Standard*, October 2001.
40. R. RIVEST, A. SHAMIR, and L. ADLEMAN, A method for obtaining digital signatures and public-key cryptosystems, *Communications of the ACM*, vol. 21, no. 2, pp. 120–126, February 1978.
41. P. FAHN and M.J.B. ROBSHAW, Results from the RSA Factoring Challenge. Technical Report TR-501, version 1.3, *RSA Laboratories*, January 1995.
42. NIST Special Publication 800-57, *Recommendation for Key Management-Part 1: General*, August, 2005.

43. NIST Special Publication 800-57, *Recommendation for Key Management-Part 2: Best Practices for Key Management Organization*, August, 2005.
44. NIST Special Publication 800-57, *Recommendation for Key Management-Part 3: Implementation-Specific Key Management Guidance*, August, 2005.
45. NIST Special Publication 800-53, *Recommended Security Controls for Federal Information Systems*, February, 2005.

Chapter 4

Cryptographic Key Distribution Systems

4.1 KEY DISTRIBUTION

Key distribution is as equally an important and sensitive operation in the cryptographic process as is cipherment and decipherment. Even the strongest cipher with believed-to-be unbreakable keys will become weak if the key distribution process is vulnerable.

How good can a cryptographic method be if the keys can be copied while in transit or during their distribution? This is a serious question, which key distribution methods attempt to address; clearly, if a method can be developed that does not require transmission of the secret deciphering key to the receiver (or to a verifying agent), and if the verifier agent can secretly create the deciphering key even without the knowledge of the rightful receiver, then a better end-to-end cryptographic process could exist to warranty message privacy and secure distribution [1]. In addition, any key distribution method should be able to revoke and distribute a new key if the already distributed key is suspected to be compromised.

Key revocation, key redistribution and new key distribution are equally important functions of the key distribution process. Key revocation is a critical operation and for this to be effective, secure monitoring mechanisms and protocols should be part of the cryptographic channel and process, or else a bad actor may, by causing continuous key revocations, cause denial of service.

The previous chapter focused on symmetric and asymmetric key cryptographic algorithms. This chapter examines the end-to-end key distribution for certain cryptographic processes: symmetric key, asymmetric public key, and shared key digital signature. Specifically, this chapter focuses on the Merkle, Shamir, Diffie-Hellman, and Digital Signatures cryptographic systems and concludes with the Trusted Third Party or Key Escrow Encryption System.

Security of Information and Communication Networks, by Stamatios V. Kartalopoulos
Copyright © 2009 Institute of Electrical and Electronics Engineers

4.1.1 Symmetric Key Distribution

As already described, the symmetric key cryptography uses the same key at both ends of the distribution cryptographic channel, at A (Alice) and at B (Bob), Figure 4.1.

Although this scheme is the simplest and the fastest in execution time, it needs to distribute the key from the point where it is generated to the other point, that is from A to B, or vice versa. Distribution of the key may be by third trusted party or by transmitting the key over a "secure" channel. The vulnerabilities in this case are the third trusted party, and the secure communication channel which may be under the attack of an eavesdropper, E (Evan); also known as the *man-in-the-middle* problem.

A solution to the trusted party problem is to use two or more trusted parties who are unknown to each other and each carrying a small piece of the key. Although this distribution method is more robust, it becomes very impractical if the key has to change very frequently. Therefore, the *more than one trusted party* distribution method is better suited to distributing scheduled secret codes (called the *schedule*), which is used over a long period of time to generate the final cryptographic key from the distributed key (Figure 4.2). Because the *schedule* contains at both ends the same seed keys that change in time, synchronization of both A and B is critical (with satellite broadcasting technology, the impact of the old synchronization problem has been greatly diminished). Notice that this method is different than public asymmetric cryptography whereby a public and a secret key are used, whereby the secret key is not the same at A and at B.

4.1.1.1 Key Revocation

If the cryptographic key is suspected to have been compromised, the cryptographic system should be able to revoke the key and distribute a new one. In the symmet-

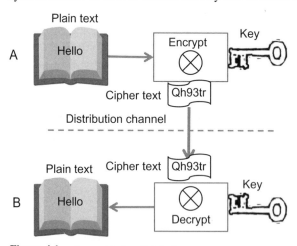

Figure 4.1. Symmetric key distribution.

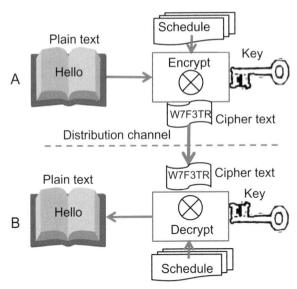

Figure 4.2.   Symmetric key distribution including a schedule.

ric case, this means that a new key should be sent to Bob, or cause a change in the schedule. Thus, the obvious question is, how does *A* and/or *B* know that the key has been compromised? A simple answer to this may be, each time Bob receives an encrypted message from *A*, he sends an encrypted receipt back to *A* (this receipt may also contain the signature of the message); this receipt is encrypted and decrypted with another key (distributed with one of the methods already described). Thus, if *A* detects a change, a key revocation protocol takes place and also a new key distribution. In this scenario, *A* or *B* may initiate the key revocation and new key distribution. In a different scenario, only *A* or *B* may have the revocation authority.

4.1.1.2   *New Key Distribution*

When a key is suspected to be compromised and has been revoked, a new key must be delivered. In this case, until the new key is delivered to Bob, Bob is unavailable to communicate securely and thus is out of contact. However, in the scheduled case, the key distribution may continue uninterrupted by simply automatically advancing the schedule and resuming secure communication. Because the final cipher key is the product of the distributed key and the seed in the schedule, which is known only to *A* and *B*, it is most unlike that the schedule is known to a third party, if it was distributed by a multi-party, as already described.

4.1.2   Asymmetric Key Distribution

Asymmetric key distribution is the process of distributing the public key in asymmetric cryptographic systems; a public key is distributed over a public channel and

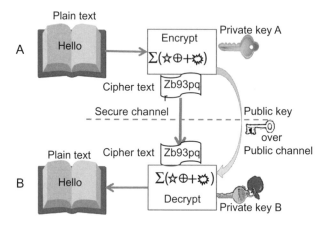

Figure 4.3. Asymmetric public key distribution.

two secret keys are used, one at *A*, which Bob does not know of, and another at *B*, which Alice does not know of. However, the two secret keys are mathematically interrelated and therefore the mathematical parameters should be distributed from *A* to *B* or vice versa before a cipher text is transmitted. As a result, distribution of the public key is not a concern as is distribution of the math parameters; the final decipher key is the math outcome of both public and private key at either end (Figure 4.3).

Public key cryptography, by virtue of the two secret keys has some interesting benefits. In digital signatures, a signed message cannot be "unsigned" by anyone else who does not have the secret key, thus proving that the message was signed by the signatory. Similarly, an encrypted message cannot be decrypted by anyone else who does not have a copy of the secret key.

In public key distribution, an important issue is the integrity of the public key since it is distributed over a public channel that anyone can access; a bad actor may capture the public key, alter it and redistribute. This issue is of serious concern and is overcome by using a third party, known as *certificate authority*, that certifies the ownership of key pairs; this is also known as *public-key infrastructure* (PKI). Thus, PKI associates a public key with its owner and it is typically accomplished by protocols [2]. However, because the key is a mathematical entity whereas the owner is not, PKI should provide explicitly statements of the association policy by means of an object identifier which is similar to an index in a catalog of registered policies.

4.1.3 Commuting Key Distribution

Assume that *A* and *B* want to exchange messages securely. However, they do not want to send keys as in symmetric or asymmetric (public) key cryptography. In this method (which is very similar to public key cryptography), Alice and Bob have their

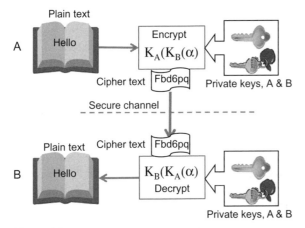

Figure 4.4. Private key sharing distribution.

own keys, K_A and K_B, respectively, which, however, are mathematically commuting. Alice and Bob have exchanged a public integer a. Alice uses her secret key to compute $K_A(a)$ and sends it to Bob, and Bob computes $K_B(a)$ and sends it to Alice. Now, Alice computes $K_A(K_B(a))$ and Bob computes $K_B(K_A(a))$. Because the keys are mathematically commuting, $K_A(K_B(a)) = K_B(K_A(a))$.

In this case, the important issue, as in public cryptography, is the integrity of the public integer a; since it is distributed over a public channel, a bad actor may capture it, alter it, and redistribute it.

4.1.4 Shared Key Distribution

The shared key is similar to symmetric key cryptography. In this case, Alice and Bob have their own private keys, although they have exchanged them. Thus A and B have both secret keys which they combine to produce a third key, the ciphering key (Figure 4.4).

Because the actual cipher key is the outcome of combining two keys, capturing KB during transmission does not infer the cipher key. However, because both keys need to be distributed, there is a possibility that both keys may be copied, unless the keys have been exchanged using a trusted agent in one direction and another trusted agent in the other direction.

4.2 MERKLE'S PUZZLE METHOD

Consider a cipher string (text) that can be deciphered in a precise amount of time. Such a string is called *puzzle* or *cryptographic primitive*. Puzzles were first introduced by Ralph Merkle whose interest started with secure communications over insecure channels [3–5].

The precision of time is calculated knowing how long is needed to decipher a puzzle. For example, if a block message is encrypted with a 40-bit key, assuming brute force, it requires 2^{40} trials to search the complete key space. If the time required for each trial is T seconds (for a given computer), and if on the average one can find it in half the trials, then it takes $2^{39} \times T$ seconds.

Now, assume that Alice and Bob have agreed on symmetric ciphers and that they can communicate only over a public channel. In addition, assume that Evan, the bad actor, eavesdrops and he has a faster computer than Alice and Bob.

Merkle's method relies on the premise that, if A and B manage to communicate in secrecy for some period (from a few hours to a few days), even if Evan will be able to eventually break the code after a period, then A and B have an advantage over Evan within this period. That is, secrecy according to this is temporary, meaning that the confidentiality of the message must also be time sensitive; an example is time critical announcements by business and governments, which declassify documents or announcements after some period. Chapter 1 we described ancient methods (such as the Polybius's square) that provided time-sensitive keys and secrecy.

How does Merkle's puzzle work?

Consider a long message. The message is partitioned in blocks of bits, the *strings*. The encrypted information starts with a block of zero bits and continues with all remaining blocks; each encrypted block is a *puzzle*. Suppose it takes one minute of computer time to decipher each of these puzzles (actual time depends on computing speed and number of instructions).

- Alice creates about 30,000 or more puzzles, each containing a pair, a 32-bit random key and a sequence number (1 to 30,000 in 32-bit binary), and transmits them to Bob in random order.

- Bob selects at random one puzzle and using the cipher key he breaks it in one minute. He keeps the key and sends back to Alice the sequence key; this also establishes a receipt that the puzzle was received by Bob.

- Alice receives the sequence number from Bob, she looks up her list of blocks or puzzles (this can be done in one simple operation with content addressable memories (CAM) [6], and she finds the corresponding cipher key. This establishes the cipher key for a lifetime T, the time it takes for Evan to break the code by brute force.

- If Evan has copied the sequence number, because he does not know the corresponding key, he must use brute force to break the code; if his computer could solve a puzzle in one millisecond, that is, 60,000 times faster than Alice or Bob, it will take Evan 2^{31} milliseconds on the average to break the code (this will take about 24 days), during which period Alice and Bob enjoy secure communication.

With modern technology and ultra-fast computers, it is possible for Evan to copy all puzzles from Alice to Bob (which contain both the key and the sequence number), listen to Bob's response with the sequence number and find the cipher key, in which case, secrecy is violated in real time. Moreover, Evan may be able to establish communication with both Alice and Bob by impersonating Alice to Bob and Bob to Alice; that is, as a man-in-the middle. Another problem with Merkle's method is its inefficiency; as the number of puzzles increases to billions, the time required by Alice to create the puzzles also increases unrealistically, and the amount of bandwidth required to transmit billions of puzzles to Bob in order to select just one, is also unrealistic. Nevertheless, it is recognized that Merkle's method set the foundation for public cryptography, since it was developed at a time when computers were slow and memory was at a premium.

4.3 SHAMIR'S KEY DISTRIBUTION METHOD

The Shamir's Key distribution method assumes that no single member is trustworthy for the whole key [7–10]. However, if the key is divided in m pieces and each piece is distributed to m members of a group, then the security of the key is increased.

The cipher key is the coefficient a_0 of a polynomial $f(x) = a_0 + a_1x + \ldots + a_{m-1}x^{m-1}$ while all other coefficients are random elements. According to Shamir's method, all m members of the group know the field of the polynomial and each one receives a point (x, y) of the polynomial. Thus, although any individual member of the group or any subgroup cannot solve Shamir's puzzle, all m can collectively determine the key. The assumption is that an eavesdropper can copy one of the key pieces but not all m.

Distribution of the m pieces over m different channels clearly enhances security if we assume that the m members of the group are not co-located. However, for the puzzle to be solved, all m members will have to send their pieces to a central point, which also is the weak distribution point.

4.4 DIFFIE-HELLMAN KEY EXCHANGE DISTRIBUTION

The Diffie-Hellman distribution system for key exchange is based on large prime numbers and modulus arithmetic. It uses a *public* large prime number, p, and a large *public* generator, α. However, in addition to this Alice has her own secret integer, $X < p$, and Bob his own secret key, $Y < p$. Because of this, such cryptographic algorithms are termed *public key* and *asymmetric key*.

The Diffie-Helman (D-H) discrete algorithm was the first published public key algorithm (Diffie and Hellman, 1976) [11], and it is based on the difficulty of solving the *discrete logarithm problem* (DLP).

DLP is briefly explained as follows:

Given a prime number p, a generator α of Zp, and a non-zero element β in Zp, find the unique integer k such that $\beta = \alpha^k \bmod p$; k is in the range from 0 to $(p-2)$ and it is called the discrete logarithm of β to the base α.

Based on this, the D-H algorithm defines a primitive root α of a prime number p such that its powers generate all integers from 1 to $p - 1$.

Thus, $\alpha \bmod p$, $\alpha^2 \bmod p$, $\alpha^3 \bmod p$, ... , $\alpha^{p-1} \bmod p$ are distinct and consist of the integers 1 through $p-1$ in some permutation.

The order of an element g (in a multiplicative group) is defined to be the smallest positive integer m such that $g^m = 1$.

An element α having order $p-1 \bmod p$ is called *primitive element mod p*. Element α is a primitive element $\bmod p$ if and only if $\alpha^{(p-1)/q} \neq 1 \bmod p$ for all primes q such that $q/(p-1)$.

As an example, consider the prime number $p = 13$. We calculate successive powers of 2 and verify that 2 is a primitive element mod 13:

$$2^0 \bmod 13 = 1, \ 2^1 \bmod 13 = 2, \ 2^2 \bmod 13 = 4, \ 2^3 \bmod e \ 13 = 8. \ 2^4 \bmod 13 = 3,$$
$$2^5 \bmod 13 = 6, \ 2^6 \bmod 13 = 12, \ 2^7 \bmod 13 = 11, \ 2^8 \bmod 13 = 9, \ 2^9 \bmod 13 = 5,$$
$$2^{10} \bmod 13 = 10, \ 2^{11} \bmod 13 = 7.$$

The element 2^i is primitive if and only if $gcd \ \{I, 12\} = 1$. Alternatively, the element 2^i is primitive if and only if $i = 1, 5, 7$ and 11, or the primitive elements modulo 13 are 2, 6, 7 or 11.

As already stated, both the primitive root α and the prime number p are the publicly known quantities in the Diffie-Hellman algorithm, whereas Alice and Bob have their own secret keys, $X < p$ and $Y < p$, respectively.

When Alice needs to communicate with Bob, she sends $K_A = \alpha^X \bmod p$ and in response Bob sends to Alice $K_B = \alpha^Y \bmod p$. Alice then computes $(\alpha^Y)^X \bmod p$ and Bob computes $(\alpha^X)^Y \bmod p$. Because α^Y and α^X are commutative, $K = (\alpha^Y)^X \bmod p = (\alpha^X)^Y \bmod p$, which is the final cipher key to be used at each end of the channel (Figure 4.5). In all this, neither Alice nor Bob know each other's integers, X and Y.

The proof is straightforward:

$$(K_B)^X \bmod p = [\alpha^Y \bmod p]^X \bmod p = [\alpha^Y \bmod p]^Y \bmod p = [K_A]^Y \bmod p$$

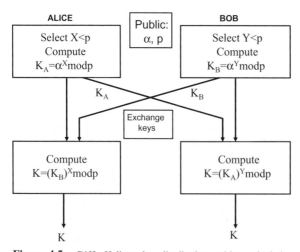

Figure 4.5.   Diffie-Hellman key distribution and key calculation.

The security of the D-H algorithm lies on maintaining the secrecy of the keys X and Y and on the difficulty to calculate discrete logarithms. For example, if Evan eavesdrops on the bi-directional channel and captures $\alpha^X \bmod p$ and $\alpha^Y \bmod p$, in order to calculate the key he must calculate Y (or X) from $\alpha^{XY} \bmod p$ using discrete algorithms, a calculation which is difficult and time consuming, particularly if the prime number is very large.

However, if Evan is successful, then he can impersonate Alice and/or Bob and as a man-in-the-middle communicate with each, without either Alice or Bob knowing it. This is overcome by using more elaborate schemes by including key authentication.

4.5 DIGITAL SIGNATURE SYSTEMS

When a document is sent in electronic form it may be necessary to have it signed, as is done with paper documents. A signature or multiple signatures (that cannot be altered) on a document, whether paper or electronic, verifies that the text was either composed by or accepted by the signatories (that is, non-repudiation), that document is authentic, and that the document was not altered when in transit (that is, text integrity). In fact, the digital signature may be considered more secure than the handwritten on paper because e-text and digital signatures are uniquely associated and are inseparable. Altering the digital signature forfeits the message validity; under certain circumstances, signatures on paper may be altered if special chemicals are used and the ink does not interact with the paper. Moreover, the digital signature verification may fail due to increased transmission errors or other reasons (such as intervention). The principles of digital signing draw from the corresponding principles of contract signing. Therefore, digital signatures should also abide by equivalent requirements, such as *fairness* (all signatories are bound equally, unless an explicit responsibility level is assigned to each signatory), *acceptance* of the document content (the protocol polls all signatories for agreement and acceptance of the document content), *completeness* and *resilience* (the protocol is resilient to adversaries who attempt to abort the signing process without the consent of the signatories), *efficiency* (the signing process is accomplished in the most secure manner, yet at minimum time and computation effort), *non-repudiation* (arbitrary withdrawal of signature is not allowed without the consent of all signatories).

There is also a case for which the recipient of the signed document signs the same document, which is returned to the document originator acknowledging that the message was received and that it was successfully deciphered. Conversely, the returned document may also indicate that the message was received with errors and was not successfully deciphered; a test of the returned document should verify the un-decipherability of the document. This proposal has not been defined by current digital signature algorithms; we call it *receipt acknowledgement with verification*. In a simplified form, this method requires a simple return receipt, which however does not indicate if the received document has errors or not. The closest to this, although not the same is a proposal by Micali, known as Micali's certified e-mail with invisible post office [12].

According to it, the protocol provides return receipts; *A* (Alice) sends to *B* (Bob) a message encrypted with the public key of the trusted third party. Bob deciphers and reads the message, and sends a receipt to Alice. Thus, Alice knows that Bob can read the message if and only if Bob sends Alice a receipt. However, if Bob fails to return a receipt, Alice may think that either Bob never received it or that the message was received in a compromised state. Clearly, this does not verify either of the two and Alice may be repeatedly sending the message in vain. Micali's protocol envisions an abort protocol in case an error was detected by Alice or by Bob. That means that Bob may have received the message and have read it and failed to reply with a receipt or failed to initiate an abort protocol.

In electronic documents, digital signatures are based on integer factorization, conventional discrete logarithms, and elliptic curve discrete logarithms [13–15]. A digital signature is represented as a binary string and is computed using a set of rules and a set of parameters associated with the identity of the signatory(ies). Because the digital signature must be uniquely associated with the signatory, an algorithm is used to generate the binary string, from an analog or digital input, using a private key. The signature is also verified but using a public key, which is mathematically interrelated with the private key but it is not the same. Thus, each user processes a private and a public key pair, and anyone can verify the signatory's signature using the publicly known key.

During the generation of the digital signature (DS), a hash function is used to create what is known as *message digest*; this is a condensed version of the text. An algorithm then receives the message digest and, using the private key, generates the digital signature. Thus, the digital signature is uniquely related to the particular text, and in this way the original text and the signature of the originator are inseparable and a third agent cannot alter the text or forge the signature.

At the receiver site, the received (message + digital signature) is verified for the originality of the message using the same hash function algorithm and the public key. The hash function is described in FIPS 180-1 [16] and digital signatures in standard PUB 186-2 [17].

According to PUB 186-2, the signature of a message *M* is the pair of the non-zero numbers *R* and *S* computed by:

$$R = \left(g^k \bmod p\right) \bmod q, \text{ and}$$
$$S = \left(k^{-1}\left(SHA - 1(M) + xR\right)\right) \bmod q$$

where *p* is a prime modulus such that $2^{L-1} < p < 2^L$ where $L = 512 + 64i$ for $0 < i < 8$, *q* divides $p - 1$ and is in the range $2^{159} < q < 2^{160}$, k^{-1} is the multiplicative inverse of *k* such that $(k^{-1}k) \bmod q = 1$, $G = H^{(p-1)/q} \bmod p$, $G > 1$, and $1 < H < p - 1$, *x* is a randomly or pseudo-randomly generated integer $(0 < x < q)$, and SHA-1(M) is the output of the SHA; a 160-bit string (SHA-1(M) is defined in PUB 186-2.

If the received versions of M, R and S are M', R', and S', the digital signature is verified by the receiver with high degree of confidence if $V = R'$, as:

$$W = (S')^{-1} \bmod q$$
$$U1 = (SHA - 1(M')W) \bmod q$$
$$U2 = ((R')W) \bmod q$$
$$V = \left(\left((G)^{U1}(Y)^{U2}\right) \bmod p\right) \bmod q$$

If $V \neq R'$, then the message is considered invalid; it is either incorrectly signed by the signatory, or it is signed by an unauthorized actor.

Approved DS algorithms for digital signature generation and verifications are the RSA (ANSI X9.31) [18, 19] and the elliptic curve digital signature algorithm (ECDSA) (ANSI X9.62).

A variant of the RSA signature is the Rabin signature scheme. With this scheme, the public key is an integer n where $n = pq$, and p and q are prime numbers which form the private key. The message to be signed must have a square root mod n; else, it must be slightly modified; about one quarter of all possible messages have square roots mod n. Thus, the signature is $s = m^{1/2}$ mod n and the verification is $m = s^2$ mod n. The Rabin signature scheme has the advantage that finding the private key and forgery are both provably as hard as factoring.

Other digital signature schemes are:

- The *blind digital signature* (the signatory signs without having knowledge of the message),

- The *one-time signature* (signature of only a single message in contrast with conventional RSA schemes for which the same key pair can be used to authenticate multiple documents). One-time signatures normally require the publishing of large amounts of data to authenticate many messages, since each signature can be used only once.

- Merkle's proposed digital signature is based on one-time signatures and on a hash function to provide an infinite tree of one-time signature. Merkle implements the signature via a tree-like scheme, where each message to be signed corresponds with a node in an arbitrarily large tree, with each node consisting of the verification parameters that are used to sign a message and to authenticate the verification parameters of subsequent nodes.

- The *any member of a group signature* allows any member of a group (with signing authority) to digitally sign a document. The receiver verifies that the signature was from an authorized member of the group but does not know which member in the group.

- The *group signature* allows members of a group to digitally sign a document. The authorized members of the group may be co-located or may not be. A verifier agent confirms that all authorized members of the group have signed or which members have not yet signed.

- The *fail-stop signature* protects against unauthorized intervention trying to forge the digital signature. It is a variation of the one-time signature scheme.
- The *undeniable signature* is a non-self-authenticating signature scheme and it uses public key cryptography based on the discrete logarithm problem. In addition to signature and verification, undeniable signature adds the *disavowal protocol*.
  - The signature part is similar to other discrete logarithm signature schemes.
  - Verification is carried out by a challenge-response protocol where the verifier sends a challenge to the signer and views the answer to verify the signature.
  - The disavowal process is also similar; A sends a challenge and B's response shows that the signature is not his. The probability that a dishonest signatory can successfully mislead the verifier in the verification or the disavowal process is $1/p$, where p is the prime number in the signer's private key. For an average 768-bit private key, the probability that the signer will be able to repudiate his signed document is negligible.

4.6 THE TRUSTED THIRD PARTY OR KEY ESCROW ENCRYPTION SYSTEM

When mailing a sensitive or important package, often one needs a confirmation that it has been received by the rightful recipient, and so requests a signed receipt. This means that the sender trusts the integrity of the carrier and believes the returned (authentic) receipt and the signature on it is authentic. That is, the sender trusts, in good faith, the professionalism and integrity of a *trusted third party*. Similarly, if one needs to sign a contract but is in a different country, the trusted third party becomes his/her embassy in that country with the authority to authenticate the signatories' signatures.

The two aforementioned examples involve paper and handwritten signatures. Now, think that all this needs to be achieved using computers. Then, one of the digital signature scenarios already described needs to be employed. However, who is the trusted third party in this case?

To assure security of the cipher text and that only the authorized user can decipher it, methods have been used that additionally provide portability of the key; that is, the authorized user, in addition to a password that allows him/her to access a server, needs a second key to encrypt/decrypt. In such system, a third trusted entity holds special data recovery master keys, which do not decrypt the cipher text but are sent to the authorized user to compute the cipher key. Because these special keys are in the trust and safeguard of a third party or escrowed, this method is known as *trusted third parties* (TTP) or *key escrow encryption system* (KEES) [20].

KEES consists of the *key escrow component*, the *user security component*, and the *data recovery component*.

- the key escrow component is managed and operated by a trusted and accountable party, an *escrow agent*, and is part of the public key certificate management infrastructure.
  - ○ Escrow agents may be private, commercial, or government; they are identified by name, location, and time-stamp (time and date).
- the user security component, a hardware device that supports the key escrow function and is capable to encrypt and decrypt.
- the data recovery component that includes the cryptographic algorithm and the communications protocol for user and device authentication.
  - ○ The cryptographic algorithm and the communications protocol in use depends on the service provider of the key encryption system. Such encryption devices may include registered portable tags, smart cards, tamper-resistant encryption chips, and PC cards. They can execute from a variety of encryption algorithms.

The security level that key escrow systems provide is commensurate with the protection against loss, detection of compromised and abused escrow devices and keys, reliability, accountability and liability of the third party, and also of the authentication of the authorized device and user.

REFERENCES

1. A. MENEZES, P.C. van OORSCHOT, and S.A. VANSTONE, *Handbook of Applied Cryptography*, The CRC Press Series On Discrete Mathematics and its Applications, CRC Press, 1997.
2. RFC 3280, R. HOUSLEY et al., *Internet X.509 Public Key Infrastructure Certificate and Certificate Revocation List (CRL) Profile*. IETF (Internet Engineering Task Force), April, 2002.
3. R.C. MERKLE, *Secure Communications over Insecure Channels*, vol 21, no. 4, Communications of the ACM, ACM Press, pp. 294–299, 1978.
4. R.C. MERKLE, "Secrecy, Authentication, and Public Key Systems", PhD Thesis, Stanford, 1979, also in UMI Research Press, 1982.
5. R.C. MERKLE, "A digital signature based on a conventional encryption function", Proceedings of Crypto'87, pp. 369–378, 1987.
6. S.V. KARTALOPOULOS, "Associative RAM-based CAM Applicable to Packet-based Broadband Communications Systems" Globecom'98, Sydney, Australia, November, 1998.
7. A. SHAMIR, How to share a secret, *Communications of ACM*, vol. 22, no. 11, pp. 612–613, 1979.
8. R.L. RIVEST, A. SHAMIR, and L.M. ADLEMAN, "Cryptographic communications system and method", U.S. Patent # 4,405,829, 20 Sept. 1983.
9. R.L. RIVEST and A.T. SHERMAN, "Randomized encryption techniques", Advances in Cryptology, Proceedings of Crypto'82, pp. 145–163, 1983.
10. R.L. RIVEST, A. SHAMIR, and D. WAGNER, "Time-lock puzzles and time-release crypto", MIT LCS Technical Report TR-684, February, 1996; also at "http://www.cs.berkeley.edu/~daw/papers/timelock.ps"
11. W, DIFFIE and M. HELLMAN, "New Directions in Cryptography", vol 22, *IEEE Transaction of Information Technology*, IEEE Computer Press, pp. 644–654, 1976.
12. S. MICALI, Certified E-mail with Invisible Post Offices, RSA Security Conference, 1997.

13. R. RIVEST, A. SHAMIR, and L. ADLEMAN, A method for obtaining digital signatures and public-key cryptosystems, *Communications of the ACM*, vol. 21, no. 2, pp. 120–126, February 1978.

14. T. ELGAMAL. A public key cryptosystem and a signature scheme based on discrete logarithms. *IEEE Transactions on Information Theory*, IT-31: 469–472, 1985.

15. FIPS PUB 186: *Digital Signature Standard*, May 19, 1994.

16. US Department of Commerce/NIST, FIPS PUB 180-1, *The Secure Hash Standard* (SHS), 2000.

17. FIPS PUB 186-2, *Digital Signature Standard* (DSS), January, 2000.

18. R. RIVEST, A. SHAMIR, and L. ADLEMAN, "A Method for Obtaining Digital Signatures and Public-Key Cryptosystems", vol 21, no. 2, in *Communications of the ACM*, ACM Press, pp. 120–126, 1978.

19. A. SHAMIR, "A fast signature scheme", MIT/LCS/TM-107, MIT Laboratory for Computer Science, 1978.

20. FIPS PUB 185, Escrowed Encryption Standard, February 9, 1994.

Chapter 5

Chaotic Cryptographic Systems

5.1 FUNDAMENTALS OF CHAOTIC PROCESSES

As the sophistication of intruders and bad actors increases, other complex processes become candidates in cryptographic systems. One of them is based on chaotic processes.

But what is chaos? Is it a disorder process? Or, is it a random process? Or, is it an undefined complex process? Because the absolute definition of chaos and its processes may be ambiguous and debatable, scientists have converged to a theory that can be mathematically described and modeled. This is the theory of chaotic processes known for short as *chaos theory* [1, 2]. Because of its complexity, chaos theory is applicable to cryptography as well as to fractals and evolutionary programming.

So, how is chaos defined in science?

In principle, chaos is the behavior of a mathematically described complex non-linear system, which has an unpredictable behavior in the following sense:

The system is extremely sensitive to initial conditions; it produces an extremely large output that resembles random noise for a very small perturbation to the initial condition, and different output for different perturbation and for different initial condition.

Because of this, chaotic systems are also known as *sensitive dependence on initial conditions*. (The term chaos in a classical sense, found in many dictionaries, has a different meaning, which may also be linked with religious beliefs; this Chapter adopts its scientific meaning applicable to cryptography).

The term *chaos* in scientific terms was popularized after 1961; in 1961 Edward Lorenz, a scientist meteorologist, working on weather simulation models found that simulation results were dramatically different when very small numerical changes were made (after the third decimal place), which were not expected to affect measurably the behavior of his model. Lorenz used the equation $X' = AX(1 - X)$, where X' is the new state, X is the previous state and A is the rate of change (a constant value

Security of Information and Communication Networks, by Stamatios V. Kartalopoulos
Copyright © 2009 Institute of Electrical and Electronics Engineers

selected to represent the problem at hand). The problem starts with some initial condition of the value X, and it proceeds to calculate the new value for X, and so on iteratively. Thus, iteratively calculating the new state from the previous one, the path of weather could be simulated from which predictions could be on the next state, thus predicting the path of a storm. The equation $X' = AX(1 - X)$ is also known as the *logistic* equation that was proposed in 1947 by Ulam and von Neumann as a source of pseudo-random numbers.

5.1.1 The Chaotic Regime

To explain how chaotic processes work, consider a linear equation $X' = AX + B$ (non-chaotic), where the initial value of $X = 0.2$, $A = 2$ and $B = -0.1$. Solving this iteratively, that is calculating $X_{n+1} = AX_n + B$, one derives a string of values for $\{X'\} = \{0.2, 0.3, 0.5, 0.9, 1.7, 3.3, \dots\}$. During this calculation, one or more parameters may change, including A or X. In the linear case, changing the initial condition from $X = 0.2$ to $X = 0.201$ and repeating the calculations, one does not expect a significant change: $\{X'\} = \{0.201, 0.302, 0.504, 0.908, 1.716, 3.32, \dots\}$.

Now, consider a non-linear equation, such as the simple second order $X' = AX(1 - X)$. Repeating the calculations for $A = 2$ and $X = 0.2$ (using the iterative $X_{n+1} = AX_n(1 - X_n)$), one calculates a string of values $\{X_{n+1}\} = \{0.2, 0.32, 0.4352, 0.49160192, 0.499858944, 0.491740606, 0.499863564, 0.499999961, 0.499999999\}$. Making a small perturbation at one of the values during the calculation process, such as $X = 0.32$ changes to $X = 0.33$, yields the results $\{0.33, 0.4422, 0.49331832, 0.49991071, 0.499999983, 0.499999999, \dots\}$, (Table 5.1).

In addition to this, other non-linear functions are used in chaos, such as $f(x,a) = (a + 1/x)^{(x/a)}$.

Plotting the values obtained for X and for various values of A on the same graph, where the horizontal axis is X and the vertical is A, a bifurcation of the graph at the perturbed point is obtained. If the coefficient A is non-linear, perturbing the coeffi-

Table 5.1. Results for $X' = AX(1 - X)$; Two left columns are before the perturbations and the two right columns after the perturbation. The perturbed value (0.33) is in bold

X	X'	X	X'
$X_0 = 0.2$	0.320000000	**$X_0 = 0.2$**	0.320000000
0.320000000	0.435200000	**0.330000000**	0.442200000
0.435200000	0.491601920	0.442200000	0.493318320
0.491601920	0.499858944	0.493318320	0.499910710
0.499858944	0.499999961	0.499910710	0.499999983
0.499999961	0.499999961	0.499999983	0.499999999
0.499999999	0.499999999	0.499999999	0.500000000

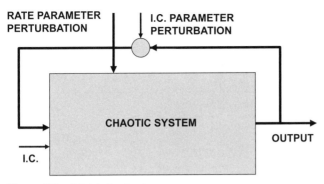

Figure 5.1. Model of a chaotic system.

Table 5.2. For $A = 2$ and initial condition (starting point) $X = 0.1$, X_n has a monotonic behavior. For $A = 3$, the results oscillate between two prongs, and for $A = 4$ the behavior is chaotic

$A = 2$	X_{n+1}	$A = 3$	X_{n+1}	$A = 4$	X_{n+1}
	$X_0 = 0.1$		$X_0 = 0.1$		$X_0 = 0.1$
	0.180000000		0.270000000		0.360000000
	0.295200000		0.197100000		0.921600000
	0.416113920		0.474754770		0.289013760
	0.485926251		0.748088035		0.821939226
	0.499603859		0.565356980		0.585420538
	0.499999686		0.737185909		0.970813325
	0.500000000		0.581229671		0.113339251
	0.500000000		0.730705221		0.401973860

cient A yields a more pronounced bifurcation, also known as the *butterfly effect*. Figure 5.1 shows the general model for a chaotic system.

Considering the iterative equation $X_{n+1} = AX_n(1 - X_n)$, it has been found that coefficient A is bounded between $0 < A < 4$ (see Figure 5.1). It has also been found that during the process, and for $0 < A < 3$ the function $A(X)$ is monotonic. Plotting A versus X yields a graph which at about $A = 3$ forks in two prongs. At some value between 3 and 4 each prong forks again. At about $A = 4$ the behavior becomes chaotic and the values oscillate over many prongs; this is known as *chaotic regime* and equations with such behavior are called *chaotic functions*. It has been calculated that at $A = 3.7$ there are 32 prongs and for $A > 3.7$ the system becomes chaotic.

Table 5.2 lists the iterative process for $A = 2$, 3 and 4 and for an initial condition (IC) $X = 0.1$. For $A = 2$, the monotonic behavior is clear, approximating 0.5 after seven iterations (0.5 is the result of truncating and rounding off after the 10th decimal place). For $A = 3$ the bifurcation is clear as the obtained values alternate between two prongs. However, for $A = 4$ the system has already fallen in chaotic oscillations

Table 5.3. The calculations in Table 5.2 are repeated but with initial condition (starting point) $X = 0.2$. For $A = 2$ and, X_n has a monotonic behavior. For $A = 3$, the results oscillate between two prongs, and for $A = 4$ the behavior is chaotic

$A = 2$	X_{n+1}	$A = 3$	X_{n+1}	$A = 4$	X_{n+1}
	$X_0 = 0.2$		$X_0 = 0.2$		$X_0 = 0.2$
	0.320000000		0.748800000		0.640000000
	0.435200000		0.564295680		0.921600000
	0.491601920		0.737598196		0.289013760
	0.499858944		0.580641290		0.821939226
	0.499999959		0.730490947		0.585420538
	0.499999999		0.590621770		0.970813325
	0.500000000		0.725363084		0.113339251
	0.500000000		0.597634440		0.401973860

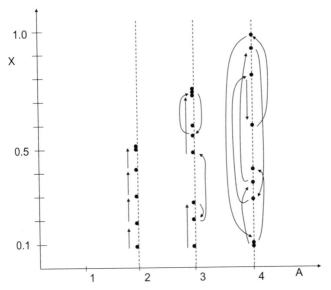

Figure 5.2.   Plotting X versus A.

among many prongs. In Table 5.3, the calculations repeat, but with an initial condition $X = 0.2$. Again, for $A = 2$, the behavior is monotonic, for $A = 3$ the bifurcation is clear, one at around 0.58 and the other around 0.74, and for $A = 4$ the system is chaotic.

The point where a system becomes chaotic is known as *point of accumulation*. Figure 5.2 plots the points of Table 5.2 for each value of A. Notice the monotonicity for $A = 2$, the oscillatory behavior about central points for $A = 3$ and the chaotic

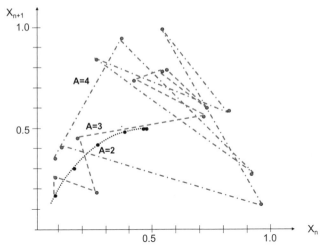

Figure 5.3.   Plots of X_{n+1} versus X_n.

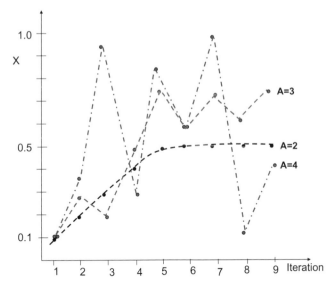

Figure 5.4.   Graphs of X versus number of iteration.

oscillatory behavior for $A = 4$. Figure 5.3 provides three scatter plots (for $A = 2$, 3 and 4) for the same data, where the vertical axis is X_{n+1} and the horizontal is X_n. Again, the monotonicity for $A = 2$, the oscillatory behavior for $A = 3$ and the chaotic behavior for $A = 4$ is obvious. Figure 5.4 plots the iteration values for the three cases. An illustration of a bifurcation diagram as proposed by R.M. May [3] for the same chaotic process is shown in Figure 5.5.

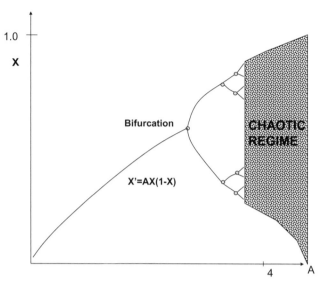

Figure 5.5.   Bifurcation map.

Table 5.4.   Some differences between Chaos cryptography and math-based algorithmic cryptography

Chaotic cryptography	Math-based algorithmic cryptography
Iterative process (closed form)	Non-iterative forward calculation
Using analysis	Using algebra and random numbers
Selectable subset of real numbers	Fixed finite set of integers
Non-integer arithmetic	Digital arithmetic
Continuous keyspace	Discrete keyspace
Uses chaotic randomn numbers	Uses pseudo-random ensembles of numbers

Based on this, one can generalize that a chaotic process is a function:

$$X_{n+1} = F(X_n)$$

The generated points, starting with an initial value of X, form a unique trajectory which enters into chaotic behavior. However, although chaotic systems exhibit random (chaotic) behavior, this chaos is reproducible, and this is what separates the scientific term of chaotic noise from pure white noise, and also chaos cryptography from math-based algorithmic cryptography (Table 5.4).

5.1.2   Chaos Annulling Traps

Table 5.5 lists results for three more initial conditions for the function $X_{n+1} = AX_n(1 - X_n)$, IC = 0.4, 0.5 and 0.6, all in the chaotic regime ($A = 4$). The results

Table 5.5.   Results for $A = 4$ and for I.C. 0.4, 0.5 and 0.6. $IC = 0.5$ causes chaos to collapse, or a CAT condition

$A = 4$	X_{n+1}	X_{n+1}	X_{n+1}	X_{n+1}	X_{n+1}	X_{n+1}
	$X_0 = 0.4$	$X_0 = 0.5$	$X_0 = 0.6$	$X_0 = 0.7$	$X_0 = 0.75$	$X_0 = 0.8$
	0.960000000	**1.0000**	0.960000000	0.840000000	**0.75000**	0.640000000
	0.153600000	**0.0000**	0.153600000	0.537600000	**0.75000**	0.921600000
	0.520028160	**0.0000**	0.520028160	0.994344960	**0.75000**	0.289013760
	0.384518112	**0.0000**	0.992596470	0.022492242	**0.75000**	0.821939226
	0.946655733	**0.0000**	0.029394870	0.087945364	**0.75000**	0.585420538
	0.201994621	**0.0000**	0.114123246	0.320843907	**0.75000**	0.970813325
	0.644771176	**0.0000**	0.404396522	0.871612376	**0.75000**	0.113339251
	0.916165225	**0.0000**	0.963439898	0.447616966	**0.75000**	0.401973860

demonstrate that even though the equation $X_{n+1} = AX_n(1 - X_n)$ is in the chaotic regime ($A = 4$), there is at least one initial condition (such as IC = 0.5) for which chaos is not only sustained but also collapses; this is an interesting result and at the parameter value for which this occurs ($A = 4$, $X_0 = 0.5$) causes a condition that is called "*chaos annulling trap*" or CAT (see corresponding column in Table 5.5 in bold).

Testing for values that are very close to $X_0 = 0.5$, $X_0 = 0.499999995$ and $X_0 = 0.500000005$, one again arrives at the same CAT condition as a result of the rounding off of the computer at hand (in fact, using a hand-held calculator, the range about 0.5 can be $X_0 = 0.4999995$ to $X_0 = 0.5000005$); IEEE recommends double precision calculations.

Similarly, for the same chaotic function and using $IC = 0.75$ the same value thereafter is derived (see Table 5.5). This condition is called "*chaos fixed trap*" or CFT (see corresponding column in Table 5.5 in bold).

When testing for values very close to 0.75, $X_0 = 0.749999999$ to $X_0 = 0.750000001$, the results oscillated about and very close to 0.75 (0.7500000002–0.749999996) but did not yield exactly 0.75.

These scenarios raise, two questions:

- For which ICs, a CAT and a CFT, are encountered in chaotic functions?
- How many iterations or cycles does it take to reach a CAT or a CFT condition?

Answers to these questions are important because the number of iterations to reach a CAT or CFT condition determines the length of the chaotic functions as a random number generator. Clearly, ICs that reach a CAT or CFT condition after many thousands of iterations are more valuable in cryptography than ICs that reach a CAT or a CFT condition in few iterations.

To determine the ICs that cause CAT or CFT conditions for chaotic functions, two methods are examined:

- An *analytic* method that starts in the chaotic regime, sets $X_{n+1} = 0$, and uses a back propagation algorithm to reach the IC at $n = 0$. However, in a general chaotic function it is not known at which iteration the CAT or CFT condition will occur, and hence what the value of n is.

- A *brute force* method that starts in the chaotic regime and tests for CAT or CFT condition for different values of X_0.

5.2 APPLICATION OF CHAOTIC SYSTEMS TO COMMUNICATIONS

Chaotic processes and systems have been applied to broadband communications for a higher level of privacy of data transmission. In telecommunications, the traditional term *narrowband* refers to the spectral band 0–4,000 Hz at a power level of approximately 0 dB that corresponds to a single telephony channel (that is, the spectral band allocated for a talker), whereas broadband refers to a band in the MHz range (as a result of multiplexing many narrowband channel). Currently, in response to a demand for higher bandwidth, the digital subscriber loop (DSL) technology is capable of transmitting broadband services to a single end-user or subscriber.

Consider that the spectrum of the narrowband (or broadband) transmitted signal has been mixed (or buried) with a chaotic random noise (Figure 5.6). Burying the signal in chaotic noise so that it is unintelligible (encrypted), transmitting it, and recovering it at the authorized destination from the mix would be an efficient data privacy method for optical and non-optical networks [4–10].

In this scenario, the receiver only should be able to exactly reproduce the same chaotic noise and subtract it from the mix {information signal + chaotic noise} to recover the original signal. However, there are two key questions:

- In the classical sense, a signal corrupted with random white noise is impossible to be "cleaned up," unless the noise spectrum has been pre-measured and analyzed. Similarly, if a signal is buried in chaotic noise, how can the receiver recover the signal?

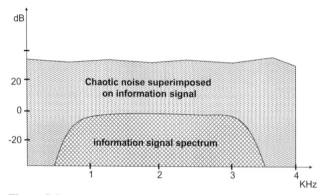

Figure 5.6.   User data is buried in chaotic noise making it completely unintelligible.

- If the signal is buried in chaotic noise, how can the two ends of the communications link be synchronized?

These questions will become clearer in the following sections.

5.3 APPLICATION OF CHAOTIC SYSTEMS TO CRYPTOGRAPHY

Chaos has demonstrated a reproducible random behavior. However, it should be pointed out that chaotic processes depend on parameter super-sensitivity, and therefore any two computers may not produce exactly the same results due to inherent accuracy that truncates numbers. That is, 16 digits accuracy (after the decimal point) may not produce the same results in a computer with 32, 64 or with 128 digits accuracy. However, for all practical purposes, this is an issue that can be easily resolved as long as one understands this ramification.

Now, the chaotic function $X_{n+1} = F(X_n)$ generates a sequence of points by iterative calculations:

$$\{X_0, F(X_0), F^2(X_0), \ldots\}$$

The point that triggers chaotic behavior is a candidate cipher key, and at the receiver, finding the key and using the same or reverse trajectory, the plain text should be derived, Figure 5.7.

5.3.1 The Baptista Method

The reproducible chaotic behavior can be used in cryptography in different ways. The Baptista method [11] uses the bifurcation or logistic map (Figure 5.5) that is generated with $X' = AX(1 - X)$.

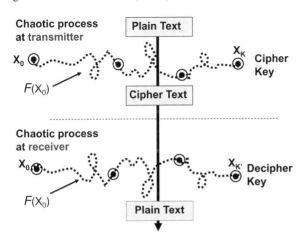

Figure 5.7. Chaotic processes at the transmitter and receiver follow two identical trajectories, according to which encipherment and decipherment are based on.

ASCII Character	!	?	A	b	.	.	.	$	#	@
S cell number	1	2	3	4	.	.	.	S-2	S-1	S

Figure 5.8.   Division of Logistic attractor into S cells and association with ASCII characters (the association may not be unique).

According to it, Baptista uses A to be large so that the system is in the chaotic regime starting with an initial condition $0 < X_0 < 1$. Then, a sequence of iterates generates points on a trajectory, and a region that is $0 < X_n < 1$ is selected. A bounded interval $X_{min} \leq X_n \leq X_{max}$ of the trajectory is divided in S cells numbered 1 to S. An ASCII character is associated with each cell (Figure 5.8). To encrypt each character of a text, one finds the number of iterations needed to reach the cell associated with that character, and the number of iterations becomes the cipher text for that character. Thus, a full cipher text consists of a string of numbers.

This process, but in reverse, is used at the receiver. The string of numbers defines the number of iterations that will point to each ASCII, thus reconstructing the text.

From the aforementioned description, one soon realizes that the Baptista method is:

- A symmetric key cryptography since the key at the transmitter and receiver are the same, and

- Nothing more than a random number generator, where each number points to a character of the ASCII alphabet. Because the frequency of occurrence of each character in the text is maintained, it presents a weakness and it makes it an easy problem to cryptoanalyze (frequency of occurrence has been a weak point to many cryptographic methods). However, since chaotic equations are many, the secrecy lies in the chaotic equation, the coefficient value, the secrecy of the min-max of the chaotic range and the cell-character association.

5.3.2   The Lorenz Method

The Lorenz method is a continuous time non-linear system. This model is based on a three dimensional model that describes a thermally-induced fluid convention in the atmosphere (Lorenz was a scientist meteorologist) with three differential (rate) equations:

$$dx/dt = \sigma(y - x)$$
$$dy/dt = \rho x - y - xz$$
$$dz/dt = xy - \beta z$$

where x is proportional to the circulatory fluid velocity, y is the temperature difference between rising-falling fluid regions, and z is the distortion of the

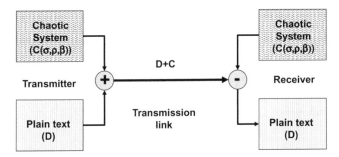

Figure 5.9. Lorenz cryptographic system.

vertical temperature profile, with initial conditions of x, y and z not necessarily the same. The parameters σ, ρ, β, in these equations are constants; σ is related to Prandtl number, ρ is related to Rayleigh number, and β is a geometric factor.

The Lorenz system of the three simultaneous differential equations leads to chaotic behavior for the values (10, 28, 8/3). Thus, these values, the initial conditions of x_0, y_0 and z_0 and a selected chaotic interval establish a cryptographic system (Figure 5.9).

Based on the Lorenz method, a similar cryptographic method has been developed [12] using a one-time pad encryption scheme, where the encryption key is determined by a chaotic sequence generated by a circuit conforming to the Lorenz system of differential equations. Synchronization of the two chaotic processes at the transmitter and the receiver is achieved using impulses. Although the one-time pad chaotic digital key enhances data privacy, the impulses that are transmitted along with the cipher text gives away an important clue to attackers and eavesdroppers. It has been demonstrated that chaotic processes can also be applied to optical communication systems [13, 14] using a chaotic laser emitter and chaos synchronization; in this method the chaotic process is applied to the optical signal using matched-pair semiconductor lasers over a commercial fiber-optic ring in downtown Athens, Greece [15]. This method is novel in the sense that the optical signal is "chaotized" to encrypt data, yet, maintaining the properties of two matched lasers at a distance of several kilometers afar is a challenge.

5.3.3 Synchronization of Chaotic Processes

The Lorenz chaotic system implies that the transmitter and the receiver at either ends of the transmission link are synchronized so that the receiver generates and removes the added chaotic noise at the transmitter.

However, one needs to consider that a pragmatic chaotic system is not as ideal as described. In fact, the receiving end does not receive a composite signal $D + C$ (Figure 5.9) but a signal with added random noise and distortions that has also suffered loss $(D + C + n)L$ (Figure 5.10); here we imply multiplication as loss is assumed

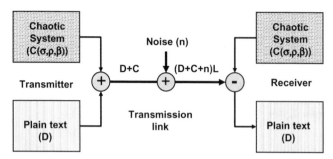

Figure 5.10.   The Lorenz Cryptographic system in pragmatic networks with random noise and signal loss.

to act on the total signal (consider that loss is the same for the bandwidth of the signal).

A ramifications of added random noise and distortion on the composite $(D + C)$ signal is that the receiver generates and subtracts the deterministic chaotic noise (C) from the composite signal $(D + C + n)L$, or $\{(D + C + n)L - C\}$, which obviously does not produce D. Thus, in chaotic communication systems, the error at the receiver is as important to consider because it determines the quality of the method; the higher the error the poorer the method.

A second ramification is synchronization of the two ends [16, 17]. Since the signal at the transmitter has been buried in chaotic noise (as part of the cryptographic process), the receiver has no means to identify the start of frame or start of packet that is typical in pragmatic transmission (circuit switched or packet switched) systems [18].

5.3.4   The Kartalopoulos Method

This section proposes a method that circumvents the synchronization issue, communicates to the receiver the ICs and other information in an encrypted manner, and provides a double encryption (classical and chaotic) to user data as follows:

1. Do not mask the complete signal with chaotic noise but leave the start of frame/packet unmasked as it is specified in traditional and next generation communication networks [19].

2. Add to the unmasked portion an embedded encrypted portion of the start of frame/packet (depending on the case, a few bits, or one or more bytes) that includes information regarding the frequency the pad should change (one-time pad, periodic pad, or fixed pad), and which pad (IC) should be used. This channel is encrypted with its own key and it does not necessitate impulses or other additional means that add to the circuit complexity and cost.

3. The user part of the data signal (D) is encrypted (E(D)) using a classical encrypted method (such as asymmetric cryptography, or synchronized symmetric), the cipher text (E(D)) is calculated to included strong error detection/correction code appended to it (such as a forward error detection/correction code, FEC), and then it is chaotically encrypted (E(D) + FEC + C).

4. The receiver first identifies the start of frame, the encrypted PAD and IC information in the added signaling (which decrypts with a different key). Based on the information in the signaling information, it removes the chaotic noise and recovers E(D) + FEC, identifies errors in the digital data stream and corrects them, and then decrypts the E(D) signal to recover the plan text D (Figure 5.11). Because the method uses the start of frame, it is self-synchronized as in the classical case, and the partially corrupted data is corrected with the FEC. Moreover, the signaling part has one key, whereas the secrecy of user data is protected by both strong classical cryptography and by chaotic processes. Finally, neither the addition of few bytes in a long packet adds to performance degradation (since it adds a very small overhead) nor the addition of FEC; FEC has already been effectively used in Optical transport networks (OTN) and in trans-oceanic networks. The amount of information added is very short, the encryption key for this portion is very small, and thus the method is very robust.

Because the method uses the start of frame, it is self-synchronized as in the classical case, and partially corrupted data is corrected with the FEC (other ECC codes may also be used).

The proposed method is backward compatible with existing standard framing protocols, which is an important aspect in pragmatic communication systems and networks, and yet it takes advantage of both classical and chaotic cryptography.

The received overall signal has been pre-amplified to compensate for transmission losses and also to match the reproduced chaotic noise level.

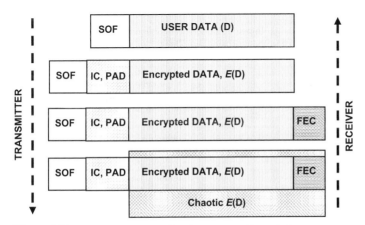

Figure 5.11. The Kartalopoulos Chaotic cryptographic method.

The user data integrity is protected by strong classical cryptography and by the chaotic process. Classical cryptography is an end-to-end path whereas the chaotic process is over the optical link (the link may also be the path).

Finally, the addition of few bytes (overhead and FEC) in a long frame/packet does not degrade the transmission performance; FEC is currently used effectively in the Optical transport network (OTN) protocol, and in trans-oceanic networks.

In summary, in the proposed method the overhead added is very short, the encryption key for this portion is very small and the method is very robust.

REFERENCES

1. J. GLEICK, *Chaos, the Making of a New Science*, Penguin Books Ltd, Harmondsworth, Middlesex, 1987.
2. ED. OTT, *Chaos in Dynamical Systems*, Cambridge University Press, 2002.
3. R.M. MAY, "Simple mathematical model with very complicated dynamics", *Nature*, vol. 261, pp. 459, 1976.
4. K.M. CUOMO, A.V. OPPENHEIM, and S.H. STROGATZ, "Synchronization of Lorenz-based chaotic circuits with applications to communications". *IEEE Trans. Circuits Syst.*, vol. 40, pp. 626–633, 1993.
5. P. COLET and R. ROY, "Digital communication with synchronized chaotic lasers", *Optical Letters*, vol. 19, pp. 2056–2058, 1994.
6. L.K. PECORA and J.L. CARROLL, "Synchronization in chaotic systems", *Phys. Review Letters*, vol. 64, pp. 821–824, 1990. REF2-SYNC.
7. A. UCHIDA, F. ROGISTER, J. GARCIA-OJALVO, and R. ROY, "Synchronization and communication with chaotic optical systems", *Prog. Opt.*, vol. 48, pp. 203–341, 2005.
8. J.P. GOLLUB and M.C. CROSS, "Chaos in space and time", *Nature*, vol. 404, pp. 710–711, 2000.
9. J. Garcia-Ojalvo and R. Roy, "Spatiotemporal communication with synchronized optical chaos", *Phys. Rev. Letters*, vol. 86, pp. 5204–5207, 2001.
10. T. MATSUURA, A. UCHIDA, and S. YOSHIMORI, "Chaotic wavelength division multiplexing for optical communications", *Opt. Letters*, vol. 29, pp. 2731–2733, 2004.
11. M.S. BAPTISTA, "Cryptography with Chaos", *Physics Letters A*, vol. 240, pp. 50–54, 1998.
12. Z. LI, K. Li, C. WEN, and Y.C. SOH, "A new chaotic secure communication system", *IEEE Trans. On Communications*, vol. 11, no. 8, pp. 1306–1312, 2003.
13. G.D. VAN WIGGEREN and R. ROY, "Communications with chaotic lasers", *Science*, vol. 279, pp. 1198–1200, 1998.
14. J.P. GOEDGEBUER, L. LARGER, and H. PORTE, "Optical cryptosystem based on synchronization of hyperchaos generated by a delayed feedback tunable laser diode", *Physical Review Letters*, vol. 80, pp. 2249–2252, 1998.
15. A. ARGYRIS, D. SYVRIDIS, L. LARGER, V. ANNOVAZZI-LODI, P. COLET, I. FISCHER, J. GARCÍA-OJALVO, C.R. MIRASSO, L. PESQUERA, and K.A. SHORE, "*Chaos-based communications at high bit rates using commercial fibre-optic links*", 26 September, 2006; correspondence and requests for materials should be addressed to Claudio R. Mirasso (claudio@galiota.uib.es).
16. P. ASHWIN, "Synchronization from chaos", *Nature*, vol. 422, pp. 384–385, 2003.
17. C.R. MIRASSO, P. COLET, and P. GARCIA-FERNANDEZ, "Synchronization of chaotic semiconductor lasers: Application to encoded communications". *IEEE Photonic Technology Letters*, vol. 8, pp. 299–301, 1996.
18. S.V. KARTALOPOULOS, *Next Generation SONET/SDH: Voice and Data*, IEEE/Wiley, 2004.
19. S.V. KARTALOPOULOS, *Next Generation Intelligent Optical Networks: From Access to Backbone*, Springer, 2007.

Chapter 6

Communication Security Layer Classifications

Previous chapters examined the mathematical foundations on which cryptographic ciphers and algorithms are based, and on which cryptographic key distribution systems are built, including chaotic cryptographic systems; Chapter 10 will also describes the foundations of quantum cryptographic systems.

Assume that one has the most difficult cipher and algorithm to break. However, as already described, this difficulty is relative. If a computer requires a long time to break the code, a faster computer requires shorter time, and a grid of computers much shorter, and serious code breakers do use superfast computers. Thus, one may assume that ciphers have a lifetime and so all ciphers fall in the puzzle classification as in Merkle's puzzle. Moreover, cipher keys are distributed and cipher texts are transmitted over conventional media and therefore they are vulnerable to copying, capturing, analyzing and attacking.

6.1 A SYNERGISTIC SECURITY FRAMEWORK

As a consequence, the best cryptography cannot warranty secrecy if the computer is not immune to attacks, and the most immune computer cannot warranty privacy if the network is not secure; that is, security is more complex than cryptography alone.

Therefore, a comprehensive cross-portfolio security framework is needed that is based on the synergy of three major security classification layers:

- the information (application and presentation),
- the computer-based, and
- the network layer.

We will refer to this synergistic cross-portfolio security framework as *network security*.

Security of Information and Communication Networks, by Stamatios V. Kartalopoulos
Copyright © 2009 Institute of Electrical and Electronics Engineers

In order to describe these three security classification layers, the Open System Interconnect (OSI) model is briefly reviewed. The OSI model defines the following seven layers (Figure 6.1).

- The two top-most layers, the application and the presentation, define means for accessing various software applications, data formats, data compression, and data encryption.

- The session layer defines functionality such as control of data between applications, grouping and data recovery.

- The transport layer defines functionality such as exchange of data between end-systems, error correction, frame sequencing, frame loss management, and quality of service features.

- The network layer defines transport of data control.

- The data link performs error detection and correction, activating and deactivating a link.

- The physical layer deals with the electrical, optical, wireless, mechanical, and functional requirements. All six layers above this layer add their own specific overhead on a data segment except the physical layer.

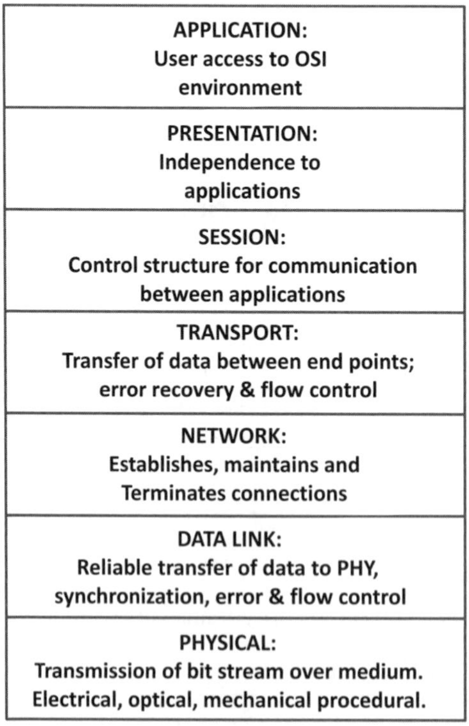

APPLICATION: **User access to OSI** **environment**
PRESENTATION: **Independence to** **applications**
SESSION: **Control structure for communication** **between applications**
TRANSPORT: **Transfer of data between end points;** **error recovery & flow control**
NETWORK: **Establishes, maintains and** **Terminates connections**
DATA LINK: **Reliable transfer of data to PHY,** **synchronization, error & flow control**
PHYSICAL: **Transmission of bit stream over medium.** **Electrical, optical, mechanical procedural.**

Figure 6.1.   Synopsis of the seven-layer OSI model.

It is important to clarify that the OSI model provides a framework of a layered organization of functionality. Although most protocols have adopted fewer than seven layers (for example, ATM has five), in principle the functionality still exists and in a similar order.

6.1.1 Information/Application Security Layer

The information or application security layer is where viruses and software attacks become an issue. This layer includes cipher key management, encryption, backup data encryption, password management, security policy and privacy architecture, malicious call trace, reporting, secure provisioning, login control, alert messaging, firewall security policy management, and more.

6.1.2 MAC/Control Security Layer

The computer-based layer or MAC/control layer is where worms may reside and cause unexpected computer behavior. Symptoms of such behavior may be:

- Clone messages, flood the network buffers, and links causing severe network congestion
- Scramble the routing tables to cause havoc
- Change the routing tables to cause traffic miss-routing
- Execute fake instructions to cause node or network shut-off or protocol misbehavior
- Others that are expected (knowledgeable bad actors are very imaginative when they want to cause mal-behavior).

6.1.3 Network Security Layer

The network layer is distinguished into several sections: network infrastructure, to security protocols (such as authentication, etc), and the physical medium. At this level is where eavesdropping takes place as well as the man-in-the-middle attacks that may disable the key distribution, cause miss-authentication, and service denial; in all, this is a critical layer that may cause secure information transmission useless.

6.1.4 Physical Layer Security

The model in this case is a simple link that includes the transmitter (the source or output port), the receiver (or input port), and the physical medium of the link. If a link contains a repeater, then depending on type of physical medium (wire, fiber, wireless) the "link" in this model is the transmitter-receiver part; a repeater in its simplest form consists of a receiver, an amplifier, filters, a re-timing unit and a transmitter). Based on this, the physical layer security is partitioned into:

a. Destination authentication

b. Source authentication

c. Link integrity monitoring
   • The destination needs to be authenticated by the source prior to transmitting secure information. Typically, this is based on a secured protocol and encrypted destination identification such as specific a signature or ID number.
   • The source needs to be authenticated by the destination prior to receiving secure information. Typically, this is based on the same secured protocol and encrypted source identification such as specific a signature or ID number.
   • The link also needs to be authenticated. Link authentication can be performed by the receiver via signal performance parameters that constitute the link signature [1, 2]; this is discussed in more detail in Chapter Seven. Additionally, the link performance needs to be continuously monitored, faults, degradations and external link attacks need to be differentiated [3], and in case of a link attack, the network (destination, source, and network management) need to be alerted and preferably execute a countermeasure scenario [4] or event a counterattack scenario [5]

6.2 FIREWALLS AND GATEWAYS

A firewall is an access control method. It consists of a communications or application gateway and provides a "liaison" function between the external Internet network and the internal Internet network. The firewall may reside on a dedicated router of the internal network that, based on per-user and per-application policies, authenticates access requests and filters incoming messages. A firewall, placed between the router of an outside (public) data network and the router of the internal (local) data network provides internal network isolation by filtering out unwanted and/or malicious packets (particularly the latter) from/to the public data network (Figure 6.2). This way, the internal router remains isolated from recognizable viruses, Trojan horses and other offensive and destructive executable programs.

The firewall router may be configured to provide security by means of access lists and access groups. An access list defines the permitted or denied actual traffic; that is, it denies connections associated with security risk and permits all others, including acceptable connections. The access list is invoked after a routing decision and before the packet is sent out. Similarly, incoming packets may be permitted for a few hosts, whereas FTP, Telnet, and rlogin may be permitted to specified hosts on the firewall subnet. Specific configuration commands are provided by router manufacturers.

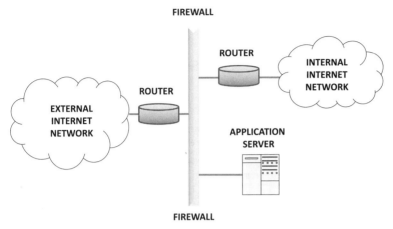

Figure 6.2.   Typical firewall architecture.

The application gateway controls the delivery of network-based services from/to the internal network, which enables users to communicate with the Internet, and/or which enables applications to establish connections between an internal and an external host.

6.3 SECURITY CROSS-PORTFOLIO

In a multivendor, multiethnic and multiprotocol network, integrity of data and security of end-to-end paths across the global network is not a simple task. For example, consider the variety of access technologies, wired, wireless and optical, each of them implemented on different protocols and having different security algorithms.

Consider also end users and large enterprise customers, the communication needs of which demand global services with predefined network availability, data deliverability, quality of service, network performance, and with data and network security. Clearly, such task requires services that can be managed and be maintained and a network implementation with additional functionality that enables security monitoring remotely, tracking of suspicious attacks, and possible countermeasures and counterattacking capabilities.

In this case, a cross-portfolio based on a security framework is necessary. Such framework includes login, password and time-outs management, secure data transport of information and across the network, secure software updates and network re-provisioning, secure network management and maintenance, malicious activity monitoring, reliable alerting, data recovery and service protection.

6.4 ATTACKS AND SECURITY IN THE INTERNET

The Internet Protocol (IP) is specifically designed for asynchronous data, which are generated and packetized by computers and are routed over a complex data network.

The nodes of this network determine the route, based on protocols and link availability at any given time (Figure 6.3).

The Internet network, which consists of data store-and-forward computer-based nodes, is particularly vulnerable to malicious attacks. At the application layer, viruses at the host node may destroy, clone or alter data. At the protocol layer, worms may gain control of the node and affect packets and routing updates. At the physical layer, external attacks may cause packet snooping, key distribution disability, service denial, and some encryption problems.

The ubiquitely used TCP/IP protocol suite [6, 7], which was developed under the sponsorship of the US Department of Defense, has been found to have a number of security flaws [8] that emanate from inadequacy of authentication mechanisms. For example, hosts rely on IP source addresses for authentication such as the Berkeley "r-utilities" [9], whereas network control mechanisms such as routing protocols have minimal authentication.

To be more specific, the *routing information protocol* (RIP) [10] is responsible for routing information on local networks, which is typically unchecked by the receiver. This may be exploited by an intruder who can send false routing information to a target host or impersonate a host, thus capturing sensitive information and passwords. If this intrusion raises the suspicion of an administrator, it may be interpreted as routing instability due to some gateway crash in the network.

Similarly, because the Internet packets suffer from high latency, and packets that travel over different routes may arrive out of order, the TCP/IP uses sequence numbers so that the destination can put the received packets in their correct order. The sequence numbers are initially established during the opening phase of a TCP connection, in a three-way handshake that is initiated with a SYN request from host *A* to host *B*. When host *B* receives the SYN request, it sends back an SYN&ACK

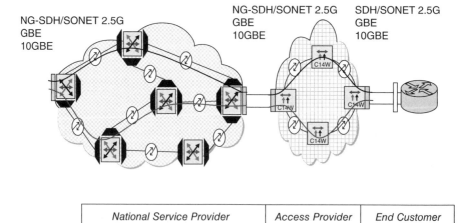

Figure 6.3.   The Internet network consists of major and moderate size network providers. Internet traffic accesses a network access server and is switched from router to router over a complex path. Each packet may be transported over a different path.

and it keeps track of the connections for a long specified period (<75 sec) to compensate for packet delay; upon receiving the SYN&ACK, host *A* replies with ACK and it completes the three way handshake. However, this handshake may be implemented differently by different hosts, which can track a different number of connections; some track 5 and some 1024. Thus, a malicious host may exploit hosts that track few connections by sending multiple SYN requests (without replying back to SYN&ACK replies to complete the three way protocol) and thus flooding the buffer of host *B* and cause denial of service; this attack is known as *SYN flooding*. In addition, when the two host are desynchronized, the attacking host may send bogus packets with correct sequence numbers in an attempt to modify the information or the network control; this is also known as *connection hijacking*.

In a different type of attack, a malicious host pretends to send packets from a different IP address than its own. The receiving IP address, because it does not authenticate the source, accepts the false packets. This is known as IP spoofing.

As a consequence, in anticipation of possible security breaches and network malicious attacks, Internet equipment manufacturers have provided features to prevent unauthorized access to routers and denial of service. Moreover, protocol standards have been defined and updated (such as, IPv4 to IPv5 to IPv6, SNMP to SNMPv2) to enhance data protection and also service including firewalls, access lists and token cards, the Simple Network Management Protocol (SNMP), the Terminal Access Controller Access Control System (TACACS) protocol, virus-detecting programs, and so on.

To combat Internet abuse or cyber-crime and provide data security, the "cyber-security" toolbox includes a plethora of protocols have been designed based on the IP security roadmap (RFC 2411) [11].

The Transport Layer Security (TLS) protocol protects peers that want to establish a secure channel in TCP-based applications [12]. TLS utilizes the X.509 identity-based security infrastructure for authentication through public key digital signatures and the corresponding public key certificates (PKC). However, TLS provides no mechanism for exchanging and negotiating authorization credentials.

In general, security in IP is addressed as follows:

- In IP-level security (IPsec) (RFC 2411), the IP authentication header (AH) (RFC 2402), and the encapsulating security payload (ESP) (RFC 2406) are in the IP datagram for authentication, integrity and confidentiality of data.
- IPsec supports the X.509 public key certificates and KeyNote assertions but it does not support security extensions. Moreover, IPsec is in the operating system kernel and thus difficult to deploy.
- The Internet key exchange (IKE) protocol (RFC 2409) is used to negotiate peer security association that consists of the cryptographic keys and the negotiated algorithms supported by the peers.

In IPv6, security features have been introduced mainly by way of two dedicated extension headers, the authentication header (AH) and the encrypted security payload (ESP).

- The AH was designed to ensure authenticity and data integrity of the IP packet. It guards against two threats: illegal modification of the fixed fields and packet spoofing.

- The ESP provides data encapsulation with encryption and data confidentiality so that only the destination node can read the payload conveyed by the IP packet. The two headers can be used together to provide all the security features simultaneously.

The security parameter index (SPI) is a parameter that specifies the security association (SA) used to decrypt and/or authenticate packets. The SPI is contained in both the AH and the ESP headers.

The AH is after the IPv6 header and the upper level payload (Figure 6.4).

The AH consists of a fixed 64-bit part and of a variable number of 32-bit blocks.

The fixed part of the AH consists of:

- The next type of payload in the chain of headers (8 bits)
- The Payload Length (8 bits); this is the total length of the authentication data in multiples of 32-bit words at a maximum of 255 32-bit blocks or 1020 bytes.
- A reserved field (16 bits)
- The SPI in the AH (32 bits)

The variable part of the AH contains the sequence number (32-bits) and the authentication data (at a maximum of 255 32-bit blocks).

When the destination node receives a packet with an AH header, the packet's authenticity and integrity is checked according to a specific procedure.

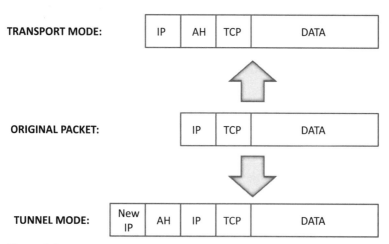

Figure 6.4. The AH is located after the IP header; shown in two modes, the transport and the tunnel mode.

The data integrity is ensured by computing and verifying the value of data according to the CRC-16 and CRC-32 algorithms. For higher data protection, the Message Digest 54 (MD54) or SHA5 algorithms may also be used in conjunction with origin authentication in order to simultaneously provide sender's identity; typically, this is accomplished using a public key encryption algorithm, which is computationally more intense than digest algorithms. The default authentication algorithm for IP security is simpler, known as keyed MD56, which is based on the keyed MD5. The keyed MD5 algorithm (that produces a 128-bit encryption key), has been shown to be attackable and thus it is expected to be replaced by a stronger encryption, such as the keyed-SHA algorithm (proposed in RFC 18527), which is based on the SHA5 (that produces a 160-bit encryption key).

The ESP is the last block in the header chain. It consists of an integer of 32-bit blocks, with the first one containing the SPI that selects the SA to be used in decrypting all other blocks in the packet. The default encryption in IPv6 is DES-CBC9, which is the DES algorithm applied in Cipher Block Chaining (CBC) mode. In the DES-CBC mode, the encrypted portion of the ESP header begins with an initialization vector composed of an integer number of 32-bit words (RFC 18299 describes the DES-CBC and RFC 185110 the 3DES-CBC that uses three different keys and is stronger than DES-CBC); DES uses a 64-bit key whereas 3DES-CBC a 112-bit key.

IP security specifies addition functionality for protocol negotiation and key exchange management. It specifies an Internet key exchange (IKE) protocol with the functionality to:

- Negotiate with other host protocols, encryption algorithms, and keys.
- Exchange keys easily.
- Keep track of all agreements.

IP security uses the SPI field in both the AH and ESP headers in order to keep track of protocol and encryption algorithm agreements. The SPI field is a 32-bit number that represents a security association (SA). The receiver assigns an available or unused SPI. It then communicates the SPI value to its communication partner establishing a SA. After this, whenever a node needs to communicate with a node having the same SA, it must use the same SPI. However, prior to communication, two nodes need to negotiate a set of keys that will be used to guarantee that the SA parameters are securely exchanged. There are two odes, the manual and the automatic.

Manual keying: the security operators have separately agreed on the keys to be used, typically at a meeting. Manual key management is applicable to restricted environments, and it requires trusted operators amd thus it is is vulnerable to human intervention.

Automatic keying: In this case, the Internet key management protocol (IKMP) is placed at the application layer, and is independent of protocols at the lower layers. The Internet security association and key management protocol (ISAKMP) defines a generic architecture for authenticated SA setup and key exchange, without specifying the actual algorithms to be used. Therefore, automatic keying can be used with different key exchange techniques.

Another proposed solution is the Oakley key-exchange protocol, which is based on a modified version of the Diffie-Hellman algorithm, and thus it is a natural partners for ISAKMP. Other proposed solutions are the Simple Key-management for Internet Protocols (SKIP), which is also based on the Diffie-Hellman algorithm. SKIP addresses several problems of key management in high-speed networks and is simple.

6.5   TACACS, SNMP and UDP

The terminal access controller access control system (TACACS) protocol allows a network access server (NAS) to offload the user administration to a central server, and it validates individual users before they gain access to the router or the communication server. In addition to the initial offering of TACAS an extension to it has been offered, called Extended TACACS or XTACACS (both TACACS and XTACACS are described in the United States DoD, RFC 1492). The protocol called TACACS+ has additional new features and it is not compatible with TACACS or XTACACS.

The SNMP protocol accesses a router and with set-request messages it configures and with get-request and get-next-request messages it gathers statistical data. SNMP packetized messages are send between a management station and the router that is equipped with an SNMP agent. Each SNMP message has a community string (a clear text password) in every packet to authenticate messages between manager and agent. The agent responds only if it receives a message with the correct community string. Because SNMP community strings are clear-text ASCII, unauthorized actors may use trap messages to capture packets from which community strings are extracted and with which router access may be gained and alter router provisioning via SNMP; this has been fixed in the second SNMP version (SNMPv2) protocol. SNMPv2 uses the MD5 algorithm (RFC 1828) to verify the communication integrity between SNMP servers and router agents (RFC 1446) by authenticating the origin and checking for timeliness. SNMPv2 can also use the data encryption standard (DES) for encrypting information.

Delay sensitive applications such as voice over IP (VoIP) prefer UDP over TCP. However, because UDP is only a mechanism for an application to send a datagram to another, it relies on the application layer for error recovery and control and therefore UDP is not a reliable datagram transport protocol and it necessitates customized security solutions. In particular, each UDP datagram is sent within a single IP datagram. IP implementations accept datagrams of 576 bytes, thus allowing for a maximum IP header of 60 bytes. The UDP has a 16-byte header and thus a UDP datagram of 516 bytes is acceptable to all IP implementations. If the IP datagram is fragmented during transmission, the receiving IP implementation re-assembles the fragments before presenting the datagram to the UDP layer.

Wrapping up, although the OSI model defines seven layers, the TCP/IP model defines five by consolidating some functionality of the seven layers into five. The five layers and the protocols in each one are:

- the Application layer with HTTP, SNMP, TELNET, DHCP, DNS, FTP, and others.
- the Transport layer with TCP, UDP, RTP, DCCP, SCTP, RSVP, and others.
- the Network/Internet layer with IPv4 and IPv6, OSPF, IPsec, ICMP, ICMPv6, and others.
- the Data Link layer with 802.11 (WLAN), 802.16 (Wi-Fi), WiMAX, ATM, DTM · Token ring, Ethernet, FDDI, Frame Relay, HDLC, PPP, ISDN, and others.
- the Physical Layer with Ethernet physical layer, Modems, PLC, SONET/SDH, OTN (G.709), physical media (Optical fiber, Coaxial cable, Twisted pair, wireless antennas), and others.

REFERENCES

1. S.V. Kartalopoulos, "Optical Channel Signature in Secure Optical Networks", *WSEAS Transactions on Communications*, Iss. 7, vol. 4, July 2005, pp. 494–504.
2. S.V. Kartalopoulos, "Real-time estimation of BER & SNR in Optical Communications Channels", *Proceedings of SPIE Noise in Communication Systems*, C.N. Georgiades and L.B. White, eds., vol. 5847, pp. 1–9, 2005.
3. S.V. Kartalopoulos, "Distinguishing between Network Intrusion and Component Degradations in Optical Systems and Networks", *WSEAS Transactions on Communications*, Iss. 9, vol. 4, pp. 1154–1161, Sept 2005.
4. S.V. Kartalopoulos, "Optical Network Security: Countermeasures in view of channel attacks", Unclassified Proceedings of Milcom 2006, October 23–25, 2006, Washington, D.C., on CD-ROM, ISBN 1–4244–0618–8, Library of Congress 2006931712, paper no. US-T-G-404.
5. S.V. Kartalopoulos, "Intelligent Optical Network Security: Counter-attack scenarios upon intrusion detection", submitted for publication.
6. E.J. Feinler, O.J. Jacobsen, M.K. Stahl, and C.A. Ward, eds. *DDN Protocol Handbook*. DDN Network Information Center, SRI International, 1985.
7. D. Comer, *Internetworking with TCP/IP: Principles, Protocols, and Architecture*. Prentice Hall, 1988.
8. S.M. Bellovin, "Security Problems in the TCP/IP Protocol Suite", Computer Communication Review, Vol. 19, No. 2, pp. 32–48, April 1989.
9. Computer Systems Research Group. *UNIX User's Reference Manual (URM). 4.3 Berkeley Software Distribution Virtual Vax-11 Version.* Computer Science Division, Department of Electrical Engineering and Computer Science, University of California, Berkeley. 1986.
10. C. Hedrick, *Routing Information Protocol.* RFC 1058, 1988.
11. S. Kent and K. Seo, *Security architecture for the internet protocol*, Draft, BBN Technologies, copyright The Internet Society, March 2005; this document with expiration date September 2005 obsoletes the same title Internet Engineering Task Force RFC 2401, 1998, by S. Kent, and R. Atkinson.
12. T. Dierks and C. Allen, RFC 2246: "*The TLS Protocol Version 1.0*", January 1999.

Chapter 7

Network Security: Wireless Systems

In a simplistic view, the network itself transports information that is generated by some end device, such as a telephone, a photographic or video camera, a computer, a sensor, and so on. The content of information should be transparent to the network and only the type of information should be of relevance to technical issues—such as data rate, synchronization, delays, and so on—that help transport information in a cost efficient manner with the expected quality of service and network performance parameters. As such, security has several facets. The first is securing the information as it is generated and prior to transmitting it, and security of the network. The first assures data privacy and as already elaborated is addressed by cryptographic methods. The second assures that the network will not be altered without appropriate clearances by unauthorized agents to copy, steal, change, or inject unauthorized information.

In addition, the network should have detectors so that if such an act is detected, it alerts the network management and also is able to execute a countermeasure strategy for self-defending or even counterattacking malicious intruders (this chapter presents such a case.) A third facet of security is the intelligent flexible network, which can be re-provisioned by remotely downloading new software versions to keep up with new features and service offerings. This remote access presents an opportunity to malicious attackers who would like to change operational and switching characteristics of nodes to either derail information or to cause havoc and denial of service. Thus, besides information security, one has to look into the network security, which often is integrated with authentication and authorization protocols, with encryption keys, and with algorithms that reside in hardware.

However, because all networks are not accessed using the same technology, we differentiate network types based on the transmission medium as each one has its own strengths and weaknesses. For example, wireless access uses electromagnetic waves that propagate in the atmosphere and in space, reaching targeted as well as untargeted antennas; in a special case, E-M waves propagate in hollow or dielectric-filled waveguides but because the waveguide medium is bounded, this case is point-

Security of Information and Communication Networks, by Stamatios V. Kartalopoulos
Copyright © 2009 Institute of Electrical and Electronics Engineers

to-point guided propagation. Electric waves propagate over a pair of metallic wires, and although electrons are guided by the wire medium, nevertheless the current flow generates a near-field magnetic field which can be detected in addition to copper being easily tapped. Finally, photons propagate in free space or in guided fiber-optic media (silica, plastic).

As a consequence, from a security point of view each accessing technology has its own personalized challenges, and encryption algorithms, key structure and protocols differ greatly. Thus each network type, based on technology, is treated separately. This chapter, examines the security of wireless networks; the next chapter examines security of optical networks.

7.1 WIRELESS NETWORKS

The transmission of electromagnetic waves is described mathematically by Maxwell's differential equations for plane waves, assuming that the observer is at some significant distance from a point source.

Wireless communications is based on transmitting and receiving modulated electromagnetic waves in full-duplex mode (receiving and transmitting simultaneously); this is in contrast to broadcasting, which is a one-way transmission (the "transmitter" does not receive and the "receiver" does not transmit) as in radio broadcasting and wireless television.

In concept, wireless communication has been a technology that has been in deployment for many decades. However, the advent of many ancillary technologies and technological factors over the last two decades has contributed to the rapid deployment of cellular communication networks and services:

- Integrated hybrid electronic circuits
- Sub-micron microelectronics
- Super dense memories
- Miniature printed circuit antennas
- High resolution displays
- Ultra-sensitive and high pixel-density photo-detecting matrices
- High energy-density (Watt-hour) rechargeable batteries
- Advanced modulation (multi-valued symbol) methods
- Advanced communication protocols
- High bandwidth optical networks
- Fusion of synchronous (telephony) and asynchronous (data) networks
- Business, smart marketing and pricing models, and
- Unprecedented user acceptance.

As a result, in just a few years, wireless telephones spread rapidly and became ubiquitous. They have been continuously shrinking and keep offering new services.

The bandwidth per user is running short; the initially few kilobits per seconds per user is now contemplated with more advanced cellular technology to megabits per second. Similarly, the few hundred users per cell is quickly exhausted and the cells are subdivided to microcells to increase the user density by frequency reuse and time-frequency division multiplexing techniques. Thus, in a manner of speaking, cellular communication has been (by definition) a disruptive technology, as it has indeed changed lifestyles.

The success of the duplex wireless telephony model triggered other wireless technologies: wireless local area networks (WLAN), wireless access (WiMax), and wireless interfaces (WiFi, Bluetooth), satellite communications (voice, data and video), each with its own specific protocol to address idiosyncratic issues and needs.

The wireless connectivity model provides users with mobility and freedom from wires which has been proven to be useful and desirable. To date, in addition to wireless cellular telephony, also available are wireless Internet connectivity, wireless mouse and keyboard connectivity, wireless peripheral connectivity, wireless control of appliances, and more.

Despite the greatness of the wireless connectivity, the electromagnetic waves reach both friendly and foe receivers. Thus, wireless communications has been vulnerable to surreptitious access, and accessing calling numbers and pin codes from the airwaves has been a relatively easy task for the connoisseur; eavesdropping, source mimicking and other unauthorized or malevolent acts have been a serious communications security and data privacy concern [1].

In an effort to improve bandwidth for new services (voice, email, internet, paging, high speed data services, image, video and global portability), to increase the mobile user density, and also to address the security concerns, cellular technology has already evolved over several generations in a relatively short time.

7.2 WLAN

Local area networks (LAN) have been in deployment for several decades; they were architected to provide a dedicated small network solution to data traffic and they were described in IEEE 802.3. Initially, they came in two sizes, small interconnecting computers within a campus (1–2 square kilometers, and medium interconnecting computers within a few square kilometers (city). Local area networks came in different topologies; for example, the Ethernet had a (wire) tree topology whereas the fiber distributed data interface (FDDI) had a dual fiber ring topology. Each LAN had its own standard protocol as well as its own advantages and disadvantages. LANs were very successful in delivering cost-efficient data and they evolved over the last three decades to deliver data services over a larger area and with more bandwidth. The current Ethernet version delivers 10 Gbps over fiber (and also over copper wire), Gbps of 100 and higher are imminent. In addition to the Ethernet, the Internet also evolved, but over a mesh topology that, however, could change and grow as nodes (routers) could be added on an ad-hoc basis by anyone; perhaps this was one of the great advantages of the Internet that helped the humongous growth of the network.

7.2.1 IEEE 802.11 WLAN

The success of wireless communications could not leave LANs untouched. Because of the mobility feature of wireless and also because a wireless interface is in almost every modern PC and laptop, wireless LAN (WLAN) was inevitable. With wireless connectivity, a stationary antenna connected to the network communicates wirelessly with a mobile antenna on the PC or other communicating appliances. Thus, computers do not need to be wire-connected to a wall but have the freedom of movement within a building, from building to building within a campus, at restaurants and cafeterias, at airports, common spaces and on the road. With multi-user protocols, a wireless computer does not have to contend for an Ethernet jack on the wall. Using spectrum (SS) or orthogonal frequency division modulation (OFDM) technology, it contends among many users for bandwidth over a designated time slot.

Currently there is a suite of wireless interface protocols under the umbrella of IEEE Standard 802.11, 1999, such as 802.11a, 802.11b, 802.11b-Corrigendum 1, 802.11d, 802.11e, 802.11F, 802.11g, 802.11h, 802.11i, 802.11j, 802.11k, 802.11m, 802.11ma, 802.11-REVma, 802.11mb, 802.11n, 802.11p, 802.11r, 802.11s, 802.11T, 802.11v, 802.11u, 802.11w, 802.11y, 802.11z, 802.11.1 and 802.11.2. Each one either describes a specific protocol according to link length (distance from antenna), transmitting power, operating frequency, access and modulation method, and data rate, or, it provides corrections and enhancements to previously published ones. Typically, WLANs operate in the 2.4 GHz S-band known as Industrial Scientific Medical (ISM) band, and in the 5 GHz public spectrum band.

7.2.2 IEEE 802.11 WLAN Architecture

The 802.11 architecture consists of the following components:

- *The wireless station (STA)* contains an adapter card, PC card, or an embedded device with an antenna for wireless connectivity.
- *The wireless access point (AP)* is wireless on the access side and wired on the network side; it acts likes a bridge between wireless STAs and the network backbone, or it bridges between one STA and another.
- *The independent basic service set (IBSS)* is a wireless network that consists of at least two STAs. It is also known as an ad hoc wireless network.
- *The basic service set (BSS)* is a wireless network that consists of a single wireless AP that supports one or more wireless users. STAs in a BSS communicate through the AP.
- *The distribution system (DS)* provides distribution services and allows STAs to roam between BSSs; it is the logical component that interconnects BSSs

with their APs; that is, APs of multiple BSSs are interconnected by the DS. This permits mobility because STAs can move from one BSS to another BSS.

- *The extended service set (ESS)* is a set of two or more APs connected to the same wired network (also known as a subnet).

Visit *www.ieee802.org/802_tutorials/nov06/802.11s_Tutorial_r5.pdf* and also *www.sss-mag.com/pdf/802_11tut.pdf* for tutorials on 802.11s.

7.2.3 IEEE 802.11 Channel Allocation

The frequency band used by 802.11 is divided into channels (or sub-bands). For example, the 2.401–2.483 GHz band is divided in 13 channels, each 22 MHz wide and with 5 MHz separation guardband; thus, channel #1 is centered at 2412 MHz and channel #13 at 2472. However, not all channels are used in every country because each one regulates channel availability differently. For example, Japan has added a 14th channel 12 MHz above the 13th channel (at the exclusion of 802.11g/n), Spain allows channels #12 and #13, whereas France only allows channels #10, #11, #12, and #13, other European countries use all 13 channels, and North America and some Central and South American countries allow all 13 channels except channels #12 and #13. Furthermore, stations can only use every fourth or fifth channel; in the Americas they use channels #1, #6, and #11, and channels #1, #5, #8 and #13 in Europe. Another small difference among different countries is the actual frequency band; in the UK it is 2400–2483 MHz whereas in the U.S. it is 2402–2483.6 MHz. The band 802.11 also specifies the permitted distribution of power across a channel. For a channel 22 MHz wide, the output signal must be attenuated by at least 30 dB at points ±11 MHz from the peak of the channel center and at least 50 dB at points ±22 MHz.

With respect to the OSI reference model, all components of the 802.11 are in the media access control (MAC) sublayer of the data link layer (DLL) or in the physical layer (PHY). The 802.11 MAC frame consists of the MAC header, the frame body, and the frame check sequence (FCS). A complete description of this is beyond the purpose of this book. It suffices, however, to identify that among the various fields in the header, the wired equivalent privacy (WEP) field indicates whether encryption and authentication are used. This field can be set for all data and management frames. The PHY sublayer defines the encoding and wireless transmission methods, DSSS, FHSS or OFDM.

The various protocols in the 808.11 suite [2–12] are (see also Table 7.1):

- The 802.11 legacy WLAN was specified at two raw data rates of 1 and 2 Mbps at a range of 20–100 meters (indoors-outdoors) specified in the ISM 2.4 GHz band (S-band).

- The 802.11a WLAN is specified in the C-band ISM 5 GHz U-NII band (actually, 5.725 to 5.875 GHz) with eight non-overlapping channels; it uses orthogonal frequency division modulation (OFDM) at a data rate of 54 Mbps,max.,

Table 7.1.   Sample characteristics of some 802.11 protocols

Protocol	Operating Frequency (GHz)	Data rate throughput (Typical) (Mbps)	Data Rate (Max) (Mbps)	Distance from antenna, indoors (meters)	Distance from antenna, outdoors (meters)
Legacy	2.4	0.9	1, 2	~20	~100
802.11a	5	23	54	~35	~120
802.11b	2.4	4.3	11	~38	~140
802.11g	2.4	19	54	~38	~140
802.11n	2.4, 5	74	248 (2 streams)	~70	~250
802.11y	3.7	23	54	~50	~5000

at a range of 35 meters (indoors); OFDM allows data to be transmitted in parallel by frequency sub-bands and provides greater resistance to interference and more throughput. When conditions are not ideal, the data rate is lowered to 48 Mbps, 36 Mbps, 24 Mbps, 18 Mbps, 12 Mbps, or 6 Mbps. Note that the frequency 5 GHz is absorbed by walls more than the 2.4 GHz.

- The 802.11b supersedes 802.11a; 802.11b is specified in the 2.4 GHz ISM band and for three non-overlapping channels. However, microwave ovens, cordless telephones and wireless garage door openers that also operate in the 2.4 GHz band and may cause interference and, potentially, denial of service. 802.11b uses direct sequence spread spectrum signaling (DSSS) and is specified at a data rate of 11 Mbps maximum for a range up to 35 meters; Bluetooth is also specified at 2.4 GHz but uses frequency hopping spread spectrum signaling (FHSS) and thus is not likely to interfere with 802.11b.

- The 802.11g supersedes 802.11b; 802.11g uses the 2.4 GHz ISM band with three non-overlapping channels; it too may suffer interference from microwave ovens and cordless telephones; it uses orthogonal frequency division modulation (OFDM) and it is specified at a higher data rate (54 Mbps) at 35 m; Bluetooth is also specified at 2.4 GHz but it uses FHSS and it is not likely to interfere with 802.11g. The 802.11g is backwards compatible with 802.11b. When the conditions are not ideal, data rates are lowered as in 802.11b.

- The 802.11i amends 802.11b to include stronger security.

- The 802.11j is an amendment specifying the WLAN Medium Access Control (MAC) and Physical Layer (PHY) in the operating band 4.9 GHz–5 GHz used in Japan.

- The 802.11n is a proposed new amendment adding multiple-input-multiple-output (MIMO) and other new features to previous 802.11 versions. The final specification is expected by June, 2009. It is specified to transmit 248 Mbps over two streams at a range of ~70 meters.

- The 802.11c-f and h are more service amendments, extensions or corrections to previous specifications.
- The 802.11y is a proposed standard operating at 3.7 GHz with 54 Mbps,max data rate at a distance 60–5000 meters (indoors-outdoors).
- The IEEE 802.11ma, 2003, merges eight amendments (802.11a,b,d,e,g,h,I,j) in a single document.
- The IEEE 802.11-2007 is the most updated document that contains all cumulative changes (up to 2007).

7.2.4 IEEE 802.11 Advantages and Disadvantages

The aforementioned plethora of WLAN specifications is the result of a successful evolution and the advantages that 802.11 offers, such as:

- *Network deployment:* The initial setup of a wireless network requires a single access point and a router connected to the data network.
- *Network expandability:* Each wireless hub serves a large number of users with compatible interface and a valid password.
- *Cost:* Wireless connectivity eliminates cables which are costly (cables + connectors + installation) if one considers that the working environment may be repartitioned and offices rearranged several times over a relatively short period (2–3 years).
- *User mobility:* a user can access the data network from anyplace, as long as there is wireless hub and the user has compatible access to the LAN.
- *User accessibility and convenience:* The wireless network allows users to access data from their primary networking environment (office) from a remote location (home, airport, road, other distant location).
- *User productivity:* Users can be productive from any place, any time when they maintain connectivity with their working environment, even if they are mobile.

Nevertheless, the plethora of WLAN specifications is also the result of 802.11 weaknesses and advantages, such as:

- *Range:* The typical range specified by 802.11g with standard equipment is tens of meters and not kilometers (at the exception of 802.11y which is specified at 5 Km and which is expected to be available in 2008).
- *Data rate:* The data rate on most wireless networks (1 to 108 Mbps) is slow compared to fast wired and particularly fiber technologies (100 Mbps to 100 Gbps).
- *Reliability:* Wireless signals are subject to fading, interference, attenuation (due to wall penetration when indoors), atmospheric phenomena (in the outdoors), and other propagation effects.

- *Security:* WLAN transceivers provide computer connectivity with a data network via electromagnetic waves that reach both legitimate and illegitimate users. Thus, it is relatively easy for a bad actor with proper equipment to intrude the network. For example, to avoid channel overlap and intrusion, WLAN uses every four or five channels with a well specified channel width that is grossly attenuated at the $\pm22\,MHz$ points from the center peak. However, bad actors within short distance (few meters) from the target user and with equipment that have shifted operating points and power above the allowed may be able to intrude. To avoid this and other intrusion types, WLANs have added authentication protocols and security algorithms.

7.2.5 WLAN Security

The weaknesses and the vulnerabilities of WLAN have been an issue of concern since its first deployment. Because of the wireless access method and because the end devices are computers, it is obvious that an intruder may access a channel, eavesdrop, plant harmful programs to the nearby computer, gain access of the network and transmit across it, and also cause unnecessary congestion.

In general, WLANs are subject to active attacks and to passive attacks. Active attacks include authentication spoofing, message injection, message modification, message decryption, virus, Trojan horse and worm planting, and the typical man in the middle. Passive attacks include cracking the encryption key, and dictionary attacks.

The IEEE 802.11, 1999 WLAN security defines the open system and shared key method for user authentication and the wired equivalency privacy (WEP) encryption algorithm for confidentiality services between two 802.11 wireless nodes.

- The open system authentication is considered a simple and null authentication algorithm; a WEP key is not necessary to authenticate a user. According to it, a station (STA) sends a request for authentication and the intended access point (AP) authenticates the client with an authentication response; that is, a station with this protocol that requests access is always authenticated if the access point is set to the open system authentication mode.

- The WEP performs data integrity by an integrity check value (ICV) in the encrypted portion of the wireless frame. WEP defines two shared encryption keys, one is the unicast session key with which unicast traffic between a wireless user and a wireless access point (AP) is protected, and the other is a multicast/global encryption key for traffic from a wireless AP to all its connected users.

The encrypted transmitted data is a combination of a 24-bit Initialization Vector (IV), the message, the RC4 symmetric stream (that uses the IV and a WEP shared key to generate the WEP seed), and the Integrity checksum for the message; the RC4 supports symmetric 40-bit and 104-bit encryption keys [13]. As a consequence, a 64-bit key and a 128-bit WEP key is available; a 128-bit WEP key is represented by a string of 26 hexadecimal numbers. For example, a 128-bit WEP seed consists of bits 0–23 corresponding to IV, and bits 24–127 corresponding to 108-bit WEP key.

Decryption follows the reverse process. Compared to the open system authentication, shared key authentication is more secure because the WEP encryption is applied during authentication. Before authentication, the WEP key is delivered to each STA via a secure channel independent of IEEE 802.11; thus, both STAs and AP hold the same WEP key.

Having the key, an STA sends an authentication request frame to AP, which responds by sending a challenge packet that contains a clear text string. The STA encrypts the challenge string with its WEP key and sends it back to AP, which decrypts it and compares with the original challenge string. If the two match, the STA is authenticated and is allowed to connect to the network. Else, the authentication fails and connectivity is denied.

Graphically, the WEP encryption function is illustrated by the block diagram, (Figure 7.1) and the encryption/decryption process (Figure 7.2).

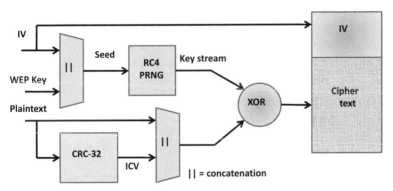

Figure 7.1. Block diagram for the WEP encipherment; "∥" indicates concatenation and "⊕" XOR (adopted from [2]).

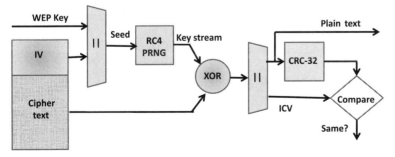

Figure 7.2. Block diagram for the WEP decipherment (adopted from [2]).

To evaluate the strengths and the weaknesses of the WEP algorithm, the following simplified description is focused on certain procedural steps.

At the transmitter:

- A CRC-32 integrity algorithm produces the Integrity Check Value (ICV) that is used for the MAC protocol data unit (MPDU).

- The seed passes the RC4 pseudo-random number generator (PRNG) and generates a key stream, which is equal to the length of the MPDU plus the ICV.

- The key sequence is XOR'ed with the data and the ICV to produce the ciphertext.

- The same IV is added to the header of the encrypted message as cleartext.

- The whole frame is transmitted. Thus, both cipher text and IV are within the same packet.

At the receiver:

- The IV is pulled out from the frame to produce a seed by combining the same symmetric WEP key.

- The WEP algorithm is executed in reverse to generate the plain text and the ICV.

- The receiver calculates the generated ICV from the plain text and compares it with the original one. If they are the same, the MPDU is correct; otherwise it is in error.

Despite these security enhancements and protocols that were included in the WEP algorithm, there have been some serious weaknesses:

- The 24-bit IV is re-used in more than one packet. If enough frames are collected, the key stream can be known and the message retrieved. Already software known as *aircrack* or *airsnort* has become available.

- User authentication is not robust and data can be delivered to unauthorized destinations.

- The WEP is an already mature protocol, and thus there is enough information and know-how for hackers to break it and invade the network.

A consequence of the WEP vulnerabilities, a more advanced 802.11 protocol was developed, known as Wi-Fi.

7.3   WI-FI, WPA AND WPA2

Before the IEEE 802.11i standard was issued, and for the sake of moving business and technology forward, the Wi-Fi Alliance created an interim solution and also a subset of 802.11i that was called Wi-Fi Protected Access (WPA) [14].

WPA includes wireless networking security features for home and small office/home office (SOHO) applications via authentication, encryption and data integrity. Data integrity is based on a message integrity code (MIC), nicknamed "Michael"; the 8-byte MIC is placed after the data field of the 802.11.

Subsequent to WPA, the 802.11i (2004) implements all mandatory elements of 802.11 and it adds MAC layer security enhancements based on the new Advanced Encryption Standard (AES), and not on the RC4 which was part of the WEP; thus, 802.11i became known as WPA2. Pursuant this, as of March 13, 2006, it is mandatory that all new Wi-Fi devices must be WPA2 certified.

Two WPA2 versions were defined: one version is defined for small office and home (SOHO) consumer applications, another is defined for enterprise applications.

The SOHO is a cost-efficient application that uses a pre-shared key (PSK) and it does not require an authentication server; thus cutting down the cost of equipment, installation and maintenance.

The enterprise application requires a WPA2 with a RADIUS server.

In addition to WPA2, the IEEE 802.11w was developed to define management and broadcast protection frames, which were not previously secured.

The 802.11i is divided into three protocols: the Temporary Key Integrity Protocol (TKIP), the Counter mode with block chaining message authentication code protocol (CCMP), and the 802.1x port access entity (PAE); a PAE can be the authenticator, the supplicant, or both.

- *The TKIP* protocol enhances the WEP privacy with a new encryption algorithm that attempt to defeat forgeries and attacks. It uses a 48-bit IV and a 48-bit key, which can be split into two phases; each phase can be calculated using different functions, thus making it difficult to an unauthorized party. WPA requires the rekeying of both unicast and global encryption keys. TKIP changes the unicast encryption key for every frame, and each change is synchronized between the wireless client and the wireless AP. For the multicast/global encryption key, WPA includes a facility for the wireless AP to advertise changes to the connected wireless clients. Thus, TKIP verifies security configuration after it determines the keys, it synchronizes the changing of the unicast encryption key for each frame, and it determines a unique starting unicast encryption key for each pre-shared authentication key.

- *The CCMP* protocol has a longer key (as long as 128 bits), which is attached to a 48-bit IV; CCMP does not have any specific hardware requirements.

- *The 802.11x* protocol is used for authentication. It is used with either encryption, TKIP or CCMP, and it is used with the RADIUS authentication server. The three components of the 802.11x framework are the supplicant, the authenticator and the authenticator server (Figure 7.3).

  ○ *A supplicant* requests access to service accessibility via the logical LAN port on a wireless LAN network adapter. The supplicant port access entity (PAE) operates algorithms and protocols that are associated with authentication. It is also responsible for communicating with the

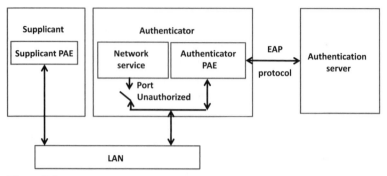

Figure 7.3. The IEEE 802.1X framework (adopted from [10]).

authenticator, and for sending the supplicant's ID when it is requested by the authenticator.

○ *An authenticator* enforces authentication prior to permitting access to service accessibility via the logical LAN port on a wireless AP. The authenticator has two PAE ports; one is uncontrolled and it communicates between the supplicant and the authentication server, and the other is the controlled port that grants or denies authorization.

○ *An authentication server* verifies the credentials of the supplicant. Authentication is accomplished according to the Extensible Authentication Protocol (EAP), an adapted point-to-point protocol (PPP) for 802.11X over LAN (EAPOL) that defines the encapsulation process for EAP messages; EAP messages are sent as payload of PPP frames.

7.4 EAP, EAPOL AND RADIUS

The Extensible Authentication Protocol (EAP) is a flexible framework that supports multiple authentication methods. Typically, EAP runs directly over data link layers such as point-to-point protocol (PPP) or IEEE 802 without requiring IP; it can also be used on dedicated links, switched circuits, wired, and wireless links [15]. There are four types of EAP message frames: EAP request, EAP response, EAP success, and EAP failure. A field in the EAP frame indicates the type frame and another field the type of request or response, including identity, MD5-challenge, and so on.

The EAP over LAN (EAPOL) defines the encapsulation method between the supplicant PAE and authenticator PAE to transport EAP frames. Five types of EAPOL packet are specified:

- EAPOL-Packet; an EAP packet is in the current EAPOL frame.
- EAPOL-Start; it starts an EAP session.
- EAPOL-Logoff; it terminates an EAP session.
- EAPOL-Key; it transmits key information between authenticator and supplicant.

- EAPOL-Encapsulated-ASF-Alert; it permits various alerts (such as specific SNMP traps) to pass through a port that is in the unauthorized state [16].

The remote authentication dial-in user service (RADIUS) protocol performs authentication, authorization, and accounting (AAA) services between a network access server (NAS) and a remote authentication server (RAS) that maintains a database of users. NAS acts as a client of RADIUS and passes user authentication information to RADIUS. Based on the information received, the RADIUS server either authenticates or does not authenticate the user, and sends to NAS one of four packets with its decision, which NAS passes to the user. RADIUS defines four types of packets: access-request, access-accept, access-reject and access-challenge [17].

Because RADIUS is extensible, many vendors create their own variants by adding extension using the vendor-specific attributes (VSA) field.

7.4.1 The IEEE 802.11i Robust Security Network

The 802.11i or WPA2 security framework introduces the robust security network (RSN) and RSN associations (RSNAs) [16]; Pre-RSN refers to 802.11 security framework(s) that were issued prior to WPA2 (Table 7.2).

The RSN supports authentication and encryption methods. The RSN information element (IE) contains an authentication and cipher suite selector that indicates different cipher suites supported by each part. Through the exchange of RSN IE information, both the STA and AP determine which cipher suit will be used. For example, STA sends an association request to AP listing the cipher suits supported by it. AP replies listing which of its cipher suit matches the list. Then both parties decide on the type of authentication and encryption algorithm to be used.

7.4.2 IEEE 802.11i Vulnerabilities

Now that WPA2 has been the mandatory standard for all new Wi-Fi equipment, the emerging debatable topic is: how secure is Wi-Fi? Are there vulnerabilities that could be exploited by bad actors?

In practice, all 802.11 protocols with radio access links are subject to malicious intervention. One that understands how these protocols work may intervene to cause at least denial of service (DoS) by disturbing the timers of the protocol. For example,

Table 7.2. Comparison between Pre-RSNA and RSNA security

	Confidentiality		Authentication		Key management
Pre-RSN	WEP		Open system	Shared key	None
RSN	TKIP	CCMP	IEEE 802.1X	PSK	4-way handshake, Group key handshake

when an algorithm is used to authenticate a user and a user sends twice unauthorized data during a one-second period, the system assumes that is under hacker attack and shuts down (this would be the action by the Michael algorithm) to protect itself. Thus, a bad actor may keep sending periodic unauthorized packets causing the system to remain out of service. Similarly, a strong signal may also jam wireless connectivity.

In general, WLAN security is very transitional in its definition and a secure protocol today may become unsecure tomorrow. The reason is that a bad actor does not have to physically connect to the network via a wire but can access traffic from any location within the range of transmission and analyze it. Although these security issues need to be addressed, another debatable topic is: how can bad actors be detected, and how can they be thwarted by the network (and not solely by the protocol)? Alternatively, how can the network self-protect using countermeasure strategies?

As a rule of thumb, the user should be aware of these security issues and risks and should use wireless access as needed and always with the latest version of a good antivirus program running on the computer. Moreover, a user should not transmit sensitive data over wireless connections and the computer should have sensitive data in files.

7.5 WIRELESS MOBILE ACCESS NETWORKS

In contrast to WLAN, which is relatively new and has focused on wireless data access, the initial wireless cellular communication network focused on wireless telephony with its deployment (and on a limited basis) several decades ago; its first appearance was made before WWII with equipment that filled a vehicle and with large antennas.

However, over the last two decades the cellular technology evolved so rapidly that today it is a multifunctional and multi-service mobile miniaturized "gadget" in almost every pocket. The drivers of the incredible shrink of the cellular phone (which is also a music player, a video camera, a video player, an internet terminal, and has large memory storage) are many: ultra-dense micro-electronics, hybrid circuitry (analog + digital) integration, high-density photodetectors, integrated antennas, high-energy density batteries, modulation methods, advanced protocols, a dense cellular network that supports synchronous and asynchronous data and smart business models that made the cellular services attractive to users. As a result, cell phones became "almost addictive" to users who once they have used it they cannot live without it; in a manner of speaking, cellular communications has changed lifestyles and thus is a revolutionary technology.

However, the details of cellular technology have not been the same over the years and for all countries, and although different protocols have been used, they too evolve to support more services and to improve security; indeed, security of cellular technology has been a serious issue from the outset. The following sections examine two representative networks, the global system for mobile (GSM) wireless communications and the third generation (3G).

7.5.1 GSM

The global system for mobile (GSM) wireless communications network, also known as the second generation (2G), enables wireless digital multiservice duplex communications with data encryption algorithms, authentication and several encrypted keys. In order to describe the level of security and the vulnerabilities, it is important to identify the key components of the network. Currently, GSM is very popular as it has been deployed in more than 200 countries [18, 19] with more than 400 million subscribers [20].

7.5.1.1 GSM Network Architecture

The GSM network architecture consists of four major functional components (text) (Figure 7.4): the mobile station (MS), the network switching system (NSS), the Base Station System (BSS) and the Operation and Support System (OSS):

- *The mobile station (MS), or cell phone*, is the handheld device that provides the service interface between end-user and wireless network. Services may be voice and data (text, image, video). The MS includes the mobile equipment (ME), the subscriber identity module (SIM), and the temporary mobile subscriber identity (TMSI) [21, 22].

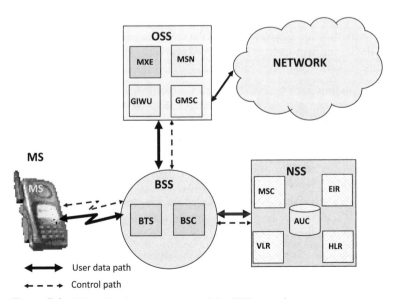

Figure 7.4. Major functional components of the GSM network.

○ Each ME is assigned at the factory a unique international mobile equipment identity (IME) number for identification purposes.
○ Each SIM contains an international mobile subscriber identity (IMSI) number that globally identifies and locates the MS. It also contains the subscriber's secret key Ki, the authentication algorithm A3, and the algorithm A8 that calculates the encryption key Kc (see below).
○ TMSI is a 32-bit temporary ID assigned to the MS by the Visitor Location Register (VLR) of the network switching system (NSS) in the serving area the MS is. During registration in a serving area, the MS sends to VLR its IMSI. The VLR, using the encryption key Kc calculates a TMSI and sends it back to the MS. As long as the MS is in the same serving area, it uses the TMSI thus making it difficult to an eavesdropper to capture the IMSI.

• *The Network Switching System (NSS)* consists of:
  ○ *The Home Location Register (HLR)*, which stores data about GSM subscribers, including Authentication Key (Ki) for each Subscriber Identity Module (SIM).
  ○ *The Equipment Identity Register (EIR)*, which contains information about the identity of mobile equipment preventing making calls from stolen, unauthorized, or defective mobile stations.
  ○ *The Visitor Location Register (VLR)* temporarily stores information about roaming GSM subscribers. Information of a roaming subscriber is copied from the exiting VLR area to the newly entered one.
  ○ *The Mobile Services Switching Center (MSC)* performs telephony switching functions and is responsible for toll ticketing, network interfacing, and common channel signaling.
  ○ *The Authentication Center (AuC)* is a database that contains the International Mobile Subscriber Identity (IMSI), the Subscriber Authentication key (Ki), and the algorithms A3 and A8 that are defined for encryption.

• *The Base Station System (BSS)* connects the base transceiver station (BTS) with the MS over a wireless (radio) interface link, and also with the OSS and NSS over a microwave, cable or fiber link. It consists of:
  ○ *The Base Station Controller (BSC)*, which provides all control functions and physical links between the MSC and BTS. It provides functions such as channel allocation, handover, cell configuration data, and control of radio frequency (RF) power levels in BTSs that are served by the BSS.
  ○ *The Base Transceiver Station (BTS)*, which handles the radio interface to the mobile station. It is the radio equipment (transceivers and antennas) that services each cell in the network.

• *The Operation and Support System (OSS)* consists of:
  ○ *The Message Center (MXE)*, which provides short message service (SMS), voice mail, fax mail, email, and paging services.
  ○ *The Mobile Service Node (MSN)*, which provides mobile intelligent network services.

- ○ *The Gateway Mobile Services Switching Center (GMSC)*, which interconnects two GSM networks.
- ○ *The GSM Interworking Unit (GIWU)*, which interfaces to various data networks.

7.5.1.2 GSM Security

From the description of registers and keys in the previous section, it is clear that GSM security uses two symmetric keys, one for authentication (Ki) and one for encryption (Kc). Security is established with the use of the following keys:

- *The Individual Subscriber Authentication Key (Ki)* is a 128-bit secret key that is shared between the Mobile Station and the Home Location Register of the subscriber's home network. Ki is generated from the SIM located in the MS utilizing the A8 algorithm (see below). For security purposes, the SIM is not accessible by the user. In the GSM, the Ki key is in the HLR and AuC registers of the NSS.

- *The Session Key (Kc)* is a 64-bit ciphering key that is used for encryption of the over-the-air channel. Kc is generated by the Mobile Station from the random challenge presented by the GSM network.

- *The Random Challenge (RAND)* is a random 128-bit stream generated by the Home Location Register.

- *The Signed Response (SRES)* is a 32-bit stream generated by the Mobile Station and the Mobile Services Switching Center.

The GSM wireless network uses three algorithms for security, the A3 for authentication, the A5 for encryption and the A8 for key generation.

- *The A3 algorithm* is used for user authentication, although authentication is an optional procedure in the beginning of a call. The A3 generates a 32-bit Signed Response (SRES), which is accomplished by utilizing the 128-bit random challenge (RAND).

- *The A5 algorithm* is used for encryption; it is a stream cipher that is initialized with the Session Key, Kc, and the number of each frame, Fn. Although the same Kc is used throughout the call, the 22-bit frame number is designed to change during the call thus creating a unique keystream for every frame. In practice, as long as the Mobile Services Switching Center (MSC) does not re-authenticate the Mobile Station, the same Kc is used for days or until the MSC re-authenticates the MS. The A5 has various encryption levels. A5/0 utilizes no encryption. A5/1 is used in Europe. A5/2 is a weaker encryption algorithm for use in the United States, and A5/3 is a strong encryption algorithm created for the 3rd Generation Partnership Project (3GPP) [23, 24].

- *The A8 algorithm* is used to generate the Kc key. From the 128-bit RAND received from the MSC, and from the 128-bit key Ki from the SIM or from the HLR, the A8 generates 128 bits, the last 64 of which are the Session Key Kc.

GSM authentication follows the steps:

1. The MS sends an authentication request to VLR for identification and its temporary mobile subscriber identity (TMSI), which was generated during registration in the serving area.

2. The VLR calculates the IMSI from the received TMSI (using the Kc) and checks its database for the subscriber's IMSI. If identified, VLR sends the authentication request to HLR; if not identified, the VLR requests the subscriber's IMSI.

3. The AuC generates the RAND and, using RAND and Ki, calculates the SRES according to the A3 algorithm. The AuC then passes RAND SRES to HLR, which sends them to VLR.

4. The VLR transmits the RAND to MS and keeps SRES.

5. The MS, with the received RAND and the key Ki, computes SRES using the A3 algorithm and sends the SRES back to VLR.

6. The VLR compares the received SRES with that which was kept; if they match, the MS is authenticated.

When the authentication process is completed, and prior to transmitting encrypted data, the MS negotiates with the network on a common *A*5 algorithm from a list of *A*5 versions. When this is done, then encrypted data can be transmitted, as soon as the keys are settled:

- The MS' SIM computes the Kc key using the *A*8 algorithm with RAND and *Ki* as input.

- The network also computes the Kc key for decryption using the same algorithm.

- Then, the MS transmits data encrypted by Kc, and the network decrypts using the same key.

7.5.1.3   GSM Full Duplex Secure Data Transmission

GSM duplex data is transmitted in a sequence of symmetric frames every 4.6 milliseconds. Each duplex frame contains 114 bits in the A to B direction, and 114 bits in the B to A. Data is encrypted; for each frame, a session key Kc is mixed with a publicly known frame counter, Fn, and the result serves as the initial state of a pseudo-random generator that produces 228 bits. These bits are XOR'ed by the two parties with the 114 + 114 bits of plain text to produce 114 + 114 bits of cipher text.

7.6 B3G/4G

The wireless cellular technology and the services offered over the cellular network evolved rapidly within a few years from a simple telephony service to an integrated voice-data-image-video-music-Internet medium, such that as soon as a next generation is announced, the next-next generation is already in the works, as is the case with GSM (or 2G), with the third generation (3G), and with the fourth generation (4G), also known as "beyond 3G" (B3G). Again, the drivers for this evolution/revolution are adding services, package miniaturization, higher data rate, and stronger security. Additionally, the wireless networking technology itself is also changing; fiber deployment is increasing, and unconditional networking methods are planned for. For example, it has been announced that the base station may change from a tall tower on which antennas and transceivers are mounted to balloons, in order to avoid real estate and installation cost; balloons with transceivers on them have already being deployed in military and other applications.

Thus, the question is: what is beyond 3G?

Because the beyond-3G standard is not yet here, several proposals are on the table. Proposals include an integration of available technologies/interworking for B3G for continuous connectivity in a heterogeneous environment [25], and a single air interface 4G at 100 Mbps when mobile, and up to 1000 Mbps when stationary [26, 27]. It has also been proposed that the MS handset simultaneously supports many access technologies to provide smooth transition from one network to another. Additionally, NTT DoCoMo introduces 4G as the mobile multimedia services—anytime, anywhere, anyone with global mobility support—and with integrated wireless solution as well as customized personal services [28].

Despite the differences in B3G proposals, all agree on ubiquitous or global access with dynamic resource provisioning and diverse and pervasive services. And, in order to achieve this, a versatile communications system is required to support customized services based on specific customer needs, commensurate with the expected quality of service (QoS), cost-efficient structure, and robust security. Additionally, a vision is that a single mobile B3G system integrates different communication technologies as cordless, cellular, WLAN, short range and wired system on a common and optimum platform which has been called Open Wireless Architecture (OWA) platform.

In the interim, the integration of 3G and WLAN is available, which is a positive step in the B3G direction [29], whereas the standardization of B3G/4G including handover and interoperability between heterogeneous networks, including 802 and non-802 networks, is in progress by IEEE 802.2 (for activities on 802.21 see www.ieee802.org/21/ and for a plenary tutorial see www.ieee802.org/21/utorials//802%2021-IEEE-Tutorial.ppt).

7.6.1 B3G/4G Security Concerns

Based on the B3G vision, the integration of diverse and heterogeneous networks raises several interoperability, compatibility and security issues.

Security becomes a serious issue as different networks offer various degrees of security, as already elaborated. In fact, customer-defined multimedia and flexible services over a diverse network will require strong and uncompromised security for information and network, as well as strategies that assure continuous and seamless mobile services even if the network is under attack.

This implies that handoff or handover is maintained at the expected QoS when the mobile device moves from one cell to the next within the same serving area, known as *horizontal handover*, or when the mobile moves from one system or network type to another (such as 2G to WLAN), known as *vertical handover*; handover from 2G to 3G has been resolved due to the similarities of the two systems.

It also implies that security at the authentication and at the encryption level should be comparable and interoperable. For example, compare security algorithms and key structure in WLAN, 3G and IP and the issues become obvious. Moreover, integration of existing heterogeneous systems require either soft or hard in-between subsystems to translate protocols from one system to the next. Although some compatibility differences may be addressed by software updates, some significant incompatibilities need to be addressed by the in-between subsystems so that from the user viewpoint the network seems to be the same. These are the challenges and more that B3G or 4G needs to resolve.

Similar to security (information and network), user privacy also becomes important. Some countries require locating the user accurately whereas some others do not [30]. However, if GPS services are to be included as part of the multimedia services offered, then locating accurately the mobile device may be unavoidable.

7.7 WIMAX

The Worldwide Interoperability for Microwave Access, commonly known as WiMax, is a broadband wireless access (BWA) solution for wireless metropolitan area networks (WMAN) and also for wireless residential access where a wired infrastructure is not feasible. That is, WiMax solves the first/last mile access problem where a wired or optical fiber infrastructure is not cost-efficient or cost-effective, while at the same time it complies with wired standards and with European BWA standards.

WiMax may also use Wi-Fi and cellular networks to provide access service as well as connectivity to the overall network. The WiMax architecture consists of subscriber stations (SS) and the base station (BS). Unlike the wireless cellular network, WiMax assumes that subscribers remain within the range the BS covers and therefore it does not does not handle handoffs. It is the next generation wireless network (B3G/4G, per IEEE 802.11e) that may utilize integrating technologies, including beam forming and MIMO technologies to offer broadband services.

WiMax offers high data rates that support both voice and video services. In residential applications, it follows the IEEE 802.16 standard at the DSL level and in enterprise applications as a fractional E1 (for Europe) or fractional T1 (for the USA).

WiMax allows up to 70 Mbps to be shared by users within a range of 30-to-50 km with line of sight and in a range of 2-to-5 km with no line of sight. The standard 802.16 defines a spectral band from 2 GHz to 60 GHz, but part of this spectral band is licence free. The WiMax standard is backed by major companies like Intel, Fujitsu, Alcatel, Siemens, BT, at&t, Qwest and McCaw [31]. If the 2.5 GHz spectrum is utilized, then companies expect to offer multiple services (voice, data, video, and GSM). The world-wide acceptance of WiMax is promising increased deployment and subscription, and many companies are on a race for getting a slice of the 2.5 GHz spectrum and also of the global market; the 2.5 GHz spectrum is crucial to the expansive success of mobile WiMax because it can be sliced to 10 MHz channels to deliver the aforementioned broadband applications.

7.7.1 WiMax Security

WiMax security, being a wireless technology, is also of serious concern. Therefore, authentication, confidentiality, privacy and assurance that the system is resistant to denial of service are all important security issues that are addressed by a sublayer placed at the bottom of the MAC layer and above the physical layer (Figure 7.5) [32, 33].

WiMax authentication is accomplished by means of a X.509 digital certificate that is issued by the manufacturer of each subscriber station (SS). The digital certificate contains the SS identity components, such as serial number, MAC address, etc. In addition, all SSs have a factory default RSA private and public key pair or they have an algorithm that generates the pair [34]. The SS initiates the process by sending an authentication request with its certificate, security capabilities, and the security association ID (SAID) to the base station (BS). The BS verifies the received certificate, generates an authentication key (AK), and sends to the SS the AK encrypted with the SS's public key, the key lifetime and the selected security suite, Figure 7.6.

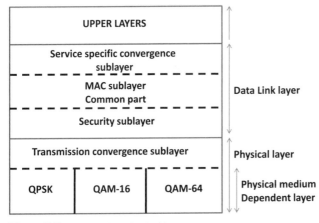

Figure 7.5. WiMax Security Architecture.

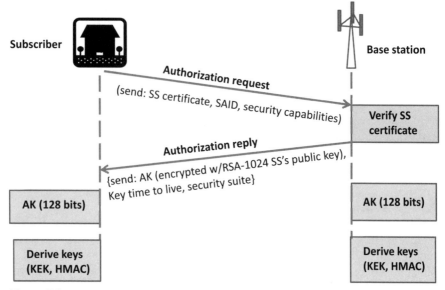

Figure 7.6.   WiMax Authentication Process.

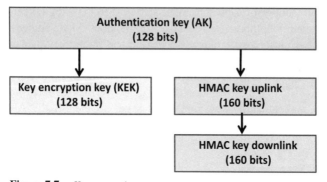

Figure 7.7.   Key generation.

As soon as the AK exchange is completed, the following keys are generated: the Key Encryption Key (KEK) is derived directly from the AK and is used to encrypt the Traffic Encryption Key (TEK), which is randomly chosen by the BS to be used for data encryption [32]. The KEK is 128 bits and is generated along with two other keys for data integrity: the HMAC for uplink and the HMAC for downlink, both 160 bit keys (Figure 7.7). The KEK key ensures confidentiality by encrypting the TEK key, which has a limited lifetime.

In conclusion, WiMax offers reasonably good wireless service and with evolving protocols, it is believed that the technology can also grow into a powerful technology with residential and SOHO high data rate applications where cables are difficult to install. However, as new services will require much higher data rates, the question is: will this wireless technology be able to meet the data rates in the access space with the service quality, service availability, and security to meet the communication needs of the next decade?

7.8 IP MULTIMEDIA SUBSYSTEMS

The IP-based Multimedia Subsystem (IMS) is an attempt to combine the quality and interoperability offered by the communications network and merge cellular mobile multi-services with the Internet. It is also an attempt to provide ubiquitous access to Internet technologies and to provide consumers with appealing multiservices worldwide. Thus, messages, computer generated data, video and voice can be shared by anyone, anywhere and anytime on any device that supports IMS.

IMS is an international standardized open architecture for mobile and fixed services. It was first specified by the 3GPP/3GPP2 and embraced by other standards bodies such as ETSI (European Telecommunications Standards Institute), TISPAN (Telecoms & Internet converged Services & Protocols for Advanced Networks), CableLabs, JCP (Java Community Process), OMA (Open Mobile Alliance) and the WiMax Forum. The IMS standard supports multiple access types, including GSM, WCDMA, CDMA2000, cable, twisted pair wire broadband access, WLAN/WiFi and WiMAX [35–38].

For IMS to be successful, the IP network and protocol should be flexible to accommodate diverse payloads, and meet the quality of service parameters and the network performance of the public telecommunications network, so that the real-time deliverability of specific payloads (such as real-time video and voice) is also met. Additionally, the IMS should be managed and be at parity in service protection and network availability with the telecommunications network.

Because the IMS network is IP-based, its architecture does not follow the traditional vertical-layered but the horizontal IP-layered structure. The Call Session Control Function (CSCF) and the Home Subscriber Server (HSS) are two major node types in IMS network architecture. The interface to the Public Switched Telephone Network (PSTN) or other wireless networks is by means of a Media Gateway Control Function (MGCF) and Media Gateway (MGW) (Figure 7.8)

- CSCF processes Session Initiation Protocol (SIP) signaling and provides session control for terminals and applications in the IMS network. The CSCF plays three roles: Serving, Interrogating and Proxy.
- HSS is the database that contains user and subscriber information, similar to Home Location Register in GSM. When a user registers in the IMS domain, the user is downloaded from the HSS to the CSCF. When more than one HSS is in the network, a Subscriber Location Function (SLF) locates the HSS that holds the subscription data for a given user.
- MGCF controls the media resources Media Gateway (MGW) when traffic flows between a Time Division Multiplexing (TDM) or synchronous network and an IP-based packet network.

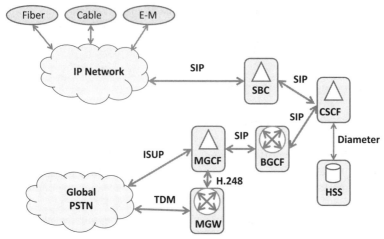

Figure 7.8. Simplified IMS network architecture and interfaces.

7.8.1 IMS Security

The main protocols used in IMS are the session initiation protocol (SIP) and the session description protocol (SDP); in addition, the AAA (authentication, authorization and accounting) protocol is used.

- SIP is the signaling protocol used in IMS (IETF, RFC 3261) and also in voice over IP (VoIP). SIP establishes, modifies and terminates multimedia sessions over IP networks, where the media delivery part is handled separately. SIP has also extensions that are designed to deliver instant message and handle subscriptions to events.

- SDP (IETF, RFC 2327) is an application-layer text-based protocol intended to describe multimedia sessions. When describing a session the caller and called party indicate their respective "receive" capabilities, media formats and receive address and port.

In IMS, users are identified with a *private user identity* and with a *public user identity*.

- *The private user identity* is a unique global identity uniquely identifies the user's subscription, and is used for authentication. The private user identity is defined by the home network operator, and is in the form of a network access identifier (NAI) such as username@operator.com (defined in RFC 2486).

- *The public user identity* is used for requesting communication with other users. Public identities can be published and take the form of a dial-up telephone number, such as +1 918-600-3240, to be reached from the PSTN, or an Internet naming such as subscriber@operator.com, to be reached over the

Internet. That is, a user can have multiple public user identities, each suitable for different purposes and convenience.

7.9 BLUETOOTH

Bluetooth is a robust, low-power and low-cost wireless technology that offers peer-to-peer communications over short distances. Bluetooth uses short-range device (SRD) radio frequency (RF) in the unlicensed 2.4-to-2.4835 GHz Industrial, Scientific and Medical (ISM) band; some European medical applications use the range 862-to-870 MHz SRD, and some U.S. applications the range 902-to-928 MHz ISM to connect with the base station located at the hospital ward where patient vitals are monitored.

Bluetooth was developed and licensed by the Bluetooth Special Interest Group (SIG), in an effort to eliminate low-level signal cables between two or more non-stationary stations that are in close proximity, such as personal computers and laptops with peripherals, mouse and keyboard, mobile phones with earphones and microphones, video game consoles, digital cameras and other appliances. Bluetooth was named after Harald Bluetooth, a late-tenth century king, by the initial special interest group (SIG) in February of 1998; it consisted of Ericcson, IBM, Intel, Nokia and Toshiba. Later, it was joined by more companies so that in ten years, by 2008, it reached more than 10,000 member-companies, sold more that 1.5 billion devices, and evolved such that in 2007 it was in version v2.1 (http://www.bluetooth.com/bluetooth/).

Bluetooth is specified by the IEEE 802.15 Wireless Personal Area Network (WPAN) standard. In WPAN applications, a wireless body-area network with sensors could provide information of vital signs such as temperature, heart rate, blood pressure, respiration, electrocardiogram (ECG) signals, skin moisture (sweat), and several others. Such sensor systems could be placed on a thin patch with all electronic components and transducers (sensors), including a microprocessor, an (ultra-thin) battery and a (printed) antenna.

The Bluetooth core system consists of an RF transceiver, baseband, and a protocol stack. Thus, Bluetooth is a wireless ad-hoc network because there is no base station and because the transceivers in the network may keep shifting their relative position. It is capable of simultaneously managing data and voice, allowing for hands-free voice calls, wireless printing, synchronizing mobile phone applications, PDAs and more.

Bluetooth uses frequency-hopping spread spectrum (FHSS) with a pseudo-random sequence known to both transmitter and receiver. FHSS reduces interference, makes bandwidth utilization more efficient and increases security. A Bluetooth transceiver hops at a rate of 1600 times-per-second over 79 channels, each channel at a nominal bandwidth of 1 MHz. The residence time at each channel is 625 μs. The Bluetooth transmission data rate is up to 1 Mbps with 720 Kbps throughput. The transmission range is from one meter up to 100 meters depending on transmitter power. Thus, depending on application, there are three classes of devices:

- Class 1 transmits at 100 mW (20 dBm) within a range up to 100 m.
- Class 2 transmits at 2.5 mW (4 dBm) within a range up to 10 m.
- Class 3 transmits at 1 mW (0 dBm) within a range up to 1 m.

Due to the ad-hoc nature of the Bluetooth network, two network structures are supported, the piconet and the scatternet (Figure 7.9).

The piconet consists of a number of devices in a star topology; however, in this topology one of the devices must assume the role of the "master device" and the others the role of the "slave devices." The master controls and sets up the network. Because a piconet has a three-bit address, up to eight devices can be active in the network, including the master. The address space of the piconet master consists of eight-bits, which means that the piconet network may include up to 255 devices; however, only eight can be active and the remaining inactive or "parked." All devices in a piconet network operate on the same channel and follow the same FHSS sequence [39]. All devices that belong to a piconet operate on the same channel and follow the same frequency hopping sequence. However, a device can belong to more than one piconets in proximity, either as a slave in both or as a master in one and a slave in another. The latter brings us to the scatternet.

The scatternet consists of two or more piconets interconnected with bridges. The bridge communicates with the piconets by synchronizing with each piconet, one at a time. Although each piconet operates in its own frequency-hopping channel, any devices in multiple piconets participate at the appropriate time through time-division multiplexing.

Bluetooth uses packet-switching and circuit-switching technologies.

An *adapted piconet channel* differs from the basic piconet channel in two ways:

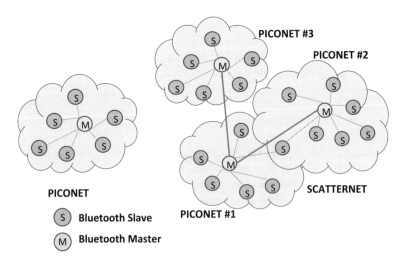

Figure 7.9.   Piconet and scatternet topologies.

- The channels on which the slaves transmit are the same as the preceding master channel. That is, the channel is not recomputed between master and subsequent slave packets.
- The adapted type can be based on fewer than 79 channels by excluding a number n ($n < 79$) of them and mark them "unused." If the pseudo-random hopping sequence selects an "unused" channel, the latter is replaced with one from the used ones ($79 - n$)

The adapted piconet channel uses the same timing and access code as the basic piconet channel. Thus, slaves in either the basic piconet channel or the adapted piconet channel are able to adjust their synchronization to the master.

7.9.1 Bluetooth Security

For Bluetooth devices to provide information protection and confidentiality from intruders, security has to be implemented at the link and application layers.

Bluetooth identifies security modes, device security levels, and service security levels as options that developers can select from.

The three security modes to select from are:

- *Security Mode 1:* non-secure; Bluetooth security functions are not used. This mode is also known as "promiscuous" since other Bluetooth devices can connect to it without authentication.
- *Security Mode 2:* service level enforced security; when the channel is established, then authentication, confidentiality and encryption take place.
- *Security Mode 3:* link level enforced security; the Bluetooth device initiates the security process before the channel is established. In this mode, authentication and encryption are carried out.

The two device security levels are:

- "trusted device" and
- "untrusted device."

The three security levels are:

- Services that require authorization and authentication.
- Services that require authentication only.
- Services that are open to all devices.

For paired devices that require user interaction, eavesdropping and the man-in-the-middle becomes an issue. Eavesdropping is addressed using a simple six-digit pass key that is claimed to be stronger than a 16-digit alphanumberic character random PIN code. This man-in-the-middle is also addressed by eliminating the possibility for an undetected middle man to intercept data.

Every Bluetooth device implements authentication and encryption the same way.

At the link layer, security is provided by using:

- a unique address for each device public (48 bits),
- a private authentication key (128 bits), also known as the link key; there are four types of authentication keys:
    - initialization key K_{init}, used for authentication during the initialization process. K_{init} is produced when two Bluetooth devices (verifier unit A and slave unit B) come across each other for the first time. K_{init} is usually derived from the Bluetooth address of the slave unit B, PIN code, PIN length in octets, and the random number (RAND) of the verifier unit A. When two devices have no previous record of a link key, authentication is accomplished by K_{init}. K_{init} is discarded after the Bluetooth devices have finished the link key exchange.
    - unit key K_A; K_A is produced in a single unit A and it depends on it.
    - combination key K_{AB}; K_{AB} depends on both units A and B and is derived from the information of both units. K_A and K_{AB} have similar function although the generating mechanism differs.
    - the master key K_{master}; K_{master} is only used for the period of the current session; it momentary replaces the original authentication key.
- a variable private encryption key (8–128 bits or 1–16 octets); it is different for different sessions and is derived from the current authentication key,
- a random number (RAND: 128 bits) that changes with every new transaction.

Despite the security features built into Bluetooth, there are still certain concerns. For example, the generation of K_{init} depends on the random number $RAND_A$ of unit A, Bluetooth device address of unit B, the PIN, and the PIN length in octets. An eavesdropper, however, can intercept $RAND_A$ in the process of transmission from Unit A to Unit B. Using an inquiry message, the hacker can obtain the Bluetooth device address of unit B. This device is now vulnerable since K_{init} depends completely on the PIN and the generation of other keys like the unit key, combination key, and master key, which also depends on K_{init}.

Having accomplished his goal, an intruder can perform one of many bad acts, such as:

- **Bluebugging:** an experienced intruder gains access to mobile phone commands without the knowledge of the owner. The intruder can then make phone calls, send and receive text messages, etc.
- **Bluesnarfing:** It is similar to Bluebugging except that it gains access to data stored on a mobile phone such as phonebook, images, calendar, etc.
- **Bluejacking:** When business cards are sent anonymously using Bluetooth, access would be gained if the recipient saves it to the phone.

• *Denial of Service (DoS):* A hacker uses a Bluetooth-enabled computer to make invalid requests for response from another device to the extent of causing temporary battery degradation and disabling the Bluetooth services.

7.10 WIRELESS PERSONAL AREA NETWORKS

Wireless Personal Area Networks (WPAN or simply PAN) is a small, local area network that consists of a computer that communicates with several peripheral devices such as personal telephones, PDAs (personal digital assistants), and other personal and wearable devices that are close to a person. Because of this, the link distance between the computer and the peripherals is typically between one to a few meters. The interface of the computer and the peripherals is wireless (such as Bluetooth) or computer-based bus (USB and FireWire).

In addition to the short links, a PAN also has a wireless link that connects with the Internet or the public communications network (including the mobile cellular or the WiMax). When the Bluetooth interface is used, then the network falls in the category of piconet with up to eight devices in a master-slave configuration as discussed in the previous section.

PAN applications range from simple telephone/PDA peripherals to more complex ones that may also include medical devices for monitoring the user's health.

Besides wireless and wired connectivity in PANs, another transmission medium is under investigation, the human body itself. The natural salinity of the human body conducts electricity, and with few nano-amperes current up to 400 Kbps can be transmitted; running a comb through hair generates more than 1000 times more current.

Using the human body model, the applications seen by IBM is passing data from one person to another or accepting (signing) an agreement by a simple handshake or touch; transfer information from a person to another will also be accomplished by touching (this makes a secure method for transferring data from person to person in confidence). Similarly business transactions will be automated and more secure. Moreover, devices with sensors could authorize a user who is wearing his/her personal PAN. Products with proper sensors would validate sale of product as soon as the buyer with a personal PAN puts it in the "sold" bin of the basket, and health data could easily be monitored by attaching sensors on the body. In short, the ingenious and imaginative painting of Michelangelo of the "touching fingers" in the Sistine Chapel to transfer encoded life to humans becomes in a manner of speaking real, particularly with the understanding of DNA and communication with the body model.

The near-field communication with the body method was developed by Thomas Zimmerman while working in MIT with magicians Penn and Teller to refine an illusion in which Penn Jillette "played" musical instruments without touching them [40–43].

Near-field communication denotes a distance less than a meter, whereas far-field denotes a distance of hundreds of meters. The fundamental difference between the two is in antenna design. Near-field communication is based on electrostatic coupling, the efficiency of which is proportional to electrode surface area. Far-field is based on antenna RF transmission, the efficiency of which depends on antenna impedance matching with that of free space; typically the antenna size is a half-wavelength. In the human body model, a transmitter in the form of a plate is in contact with the body and it couples the electric field of the signal to the body. A modulated electric current flows through the body and reaches other receivers connected to the body, like sensors of an electro-cardiogram (ECG) machine; receiver sensors are grounded so that any change in the body electric field is with respect to ground, or 0 Volts. Thus, the signal that is received is amplified, filtered from noise sources, and reshaped into the signal as it was initially transmitted. The decoder converts it to digital data which is processed as needed. The PAN is based on the seven layer ISO 7498 model. Two modulation methods have been evaluated, the ON–OFF keying and direct sequence spread spectrum modulation (DSSSM) that uses a pseudo-random noise sequence. The 802.15 working group is defining different versions for devices that have different requirements. For example, 802.15.3 deals with high-bandwidth (about 55 M bps), low-power MAC and physical layers. Section 802.15.4 deals with low-bandwidth (about 250 K bit/sec), extra-low power MAC and physical layers.

Assuming wireless PAN devices that have a length between 25 mm (wristwatch) to 80 mm (credit card), several GHz would be required for efficient transmission. Because the energy dissipation increases with frequency, the maximum carrier frequency is critical to near-field battery operated devices. Therefore, near-field devices operate at low frequencies between 0.1 to 1 MHz. Because peripherals contain many similar or same functions, integration of peripherals will remove function redundancy, which leads to miniaturization and an all-in-one peripheral with lower power consumption.

7.10.1   Wireless Personal Area Networks Security

As with all wireless end-devices, WPAN devices periodically transmit a unique user code to a stationary transceiver for identification, location, and authorization so that data exchange can take place; this is in addition to the method (wireless or body contact to connect peripherals to the personal/body computer). Although the transmitted signal is very weak compared to far-field devices, privacy becomes important as bad actors with a sensitive receiver and a stronger transmitter may intervene the near-field signal and perform one of the acts described previously (eavesdrop, disrupt, inject, alter, etc.); notice that near-field transceivers communicate, in addition to voice, music, and video, personal and sensitive information like credit card numbers, banking data, etc. Similarly, when using the body model, touching a body or a telephone wire has the same effect; eavesdropping can take place as easily. That is, as with any communication system, PANs are also confronted with the same security challenges.

Therefore, in addition to the transmission method (FDMA, TDMA, CDMA), secure protocols and encryption algorithms are needed, which, however, increases the processing power of the local processor and thus the power consumption.

7.11 COMMUNICATIONS SATELLITE NETWORKS

A satellite travels on orbits around the earth and outside the atmosphere at various speeds and different heights. Orbits are classified according to their altitude as a low orbit, medium orbit, or geostationary orbit. Direct orbit is the one in which a satellite moves in the direction the Earth moves (spins). A retrograde orbit is the one in which the satellite moves in the opposite direction from which the Earth moves. An inclined orbit is the one in which the satellite moves in an angle with respect to the equator. Low earth orbit satellites (LEOS) travel incline orbits and at a height approximately 500 Km. GPS satellites orbit at approximately 20,200 Km which is known as medium earth orbit (MEO). Geostationary satellites orbit the earth at 42,300 Km; geostationary satellites complete a revolution of the Earth in 24 hours, and therefore they seem to be stationary (hence their name). For example, for a geostationary satellite with a height of approximately 35,800 Km, the satellite's speed is 11,000 Kmph for a period T = 24 hrs (Figure 7.10).

Satellites can be as simple as transponder satellites or as switching satellites.

- Transponders act as regenerators: they receive a signal, retime, re-shape and amplify. Additionally, they may change the carrier frequency from the up-link to downlink (Figure 7.11). A transponder satellite communicates with two earth stations, approximately 6500 Km apart, via an uplink frequency (<40 GHz) and a downlink frequency (<20 GHz). The transmit power of an earth station is approximately 200 dBW.

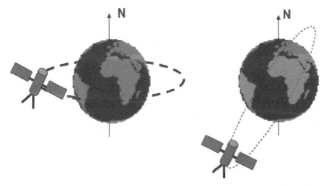

Figure 7.10. Geosynchronous satellites are positioned on orbits at the regions plane whereas the MEOS and LEOS on inclined orbits to cover the polar.

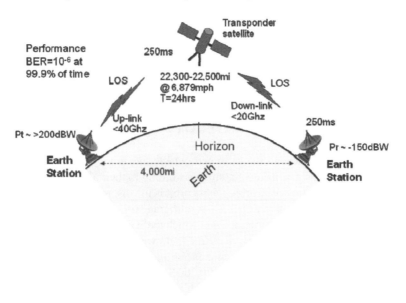

Figure 7.11.   Earth station and geosynchronous satellite system.

- Switching satellites act as switching communication nodes; that is, nodes in the sky. They have several input and output ports and in addition to the regenerator function, they also switch (or re-direct) a signal from one output port to another. Clearly, this is a more complex satellite system as it requires more computational and hardware intelligence, more intelligent protocols, more antennas and more complex power generation and management [44–45], (Figure 7.12). (Antennas in Figure are denoted N, S, E and W, in addition to up-link/down-link.)

Asynchronous satellites (LEOS and GPS) roam the sky in constellations; the LEOS Teledesic network was initially planned for 840 satellites, the LEOS Iridium for 66 satellites and the MEO GPS for 30 satellites on-orbit, with 24 satellites being active and six as spares and ready to replace any failing satellite out of the 24 active.

Typically, satellite constellations form a mesh network and therefore in addition to antennas for traffic up-linking/down-linking from/to terrestrial stations, they include antennas or laser transceivers for inter-satellite links (ISL) as in Figure 7.12, or cross-linking, [44]; the photo of Figure 7.13 depicts a commercial satellite with multiple antennas and Figure 7.14 shows the concept of a satellite communications network with laser ISLs. Such networks may be or may not be geostationary.

Satellites, whether synchronous or asynchronous, provide different communication services, such as voice, video, data, ATM and IP, broadcast or two-way, with the objective to complement (and not to compete with) transoceanic submarine fiber

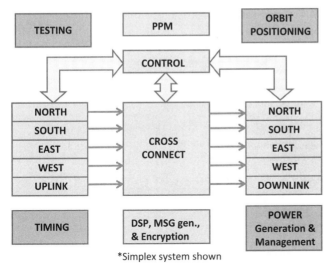

Figure 7.12. Functional block of a switching communication satellite with ISIs.

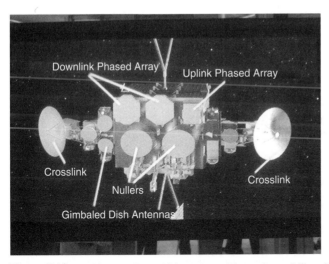

Figure 7.13. A satellite with multiple antennas (photo taken at Milcom'06).

optic networks (Figure 7.15). The GPS network offers commercial services for laptops, cars, ships, airplanes, portable devices equipped with low cost (trans)receivers, as well as video broadcasting; the latter is gaining vast popularity in many countries where video cable is not easily deployable or where the cable does not support certain channels that appeal to subscribers. It also offers military services (such as, NavWar in the U.S. and Galileo in Europe) for portable devices with anti-jamming, security, jamming/anti-jamming, and more.

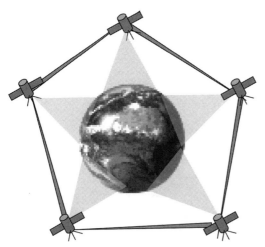

Figure 7.14. A satellite network multiple laser ISLs.

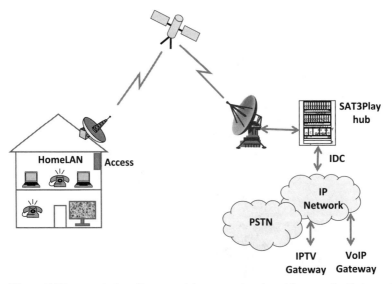

Figure 7.15. A typical satellite network is connected to the public network offering various synchronous and asynchronous services.

A satellite network encompasses three major segments:

- The space segment comprises all airborne satellites in orbit. Each satellite has a payload and a platform. The payload includes all receiving and transmitting antennas and any additional equipment for antenna control.

- The ground segment is the terrestrial receiving/transmitting station. It consists of the antenna and interface equipment that connect to the terrestrial network or to an end-user communications device. Depending on application, an antenna may be from under 0.6 meters in diameter and up to 30 meters. Small antennas are typically used in mobile or portable applications and large antennas are used in stationary ground station applications [46, 47].

- The control segment on terrestrial facilities orchestrates, manages, supervises and monitors transmission between satellites in space and ground stations. There are two types of control:
  - One type of control is responsible for satellite tracking, telemetry, and command (TTC).
  - The other type manages and controls the onboard resources of the satellite.

Typically, one of three modulation methods are adopted for the uplink/downlink beam of commercial satellites, frequency division multiple access (FDMA), time division multiple access (TDMA), or code division multiple access (CDMA).

- In FDMA, the bandwidth is divided into bands with sufficient guard band between them and each earth station is assigned one or more bands such that two adjacent stations cannot interfere.
- In CDMA, a unique orthogonal code sequence is assigned to each station which is used to modulate the signal, thus ensuring that each station sends a uniquely modulated signal. This method is also used in wireless cellular systems because of many advantages, such as:
  - Frequency-dependent transmission impairments such as noise bursts and selective fading have less effect.
  - More resistance to multipath impairments.
  - Better inherent privacy.
  - As more users simultaneously access the system, the noise level and error rate gradually increases.
- In TDMA each earth station is allocated a time slot of fixed duration within which each station transmits/receives. Therefore, TDMA is less vulnerable to intermodulation and efficiently uses the transmitting power (transmitter is active only during its time slot). TDMA also allows for digital switching (telecommunications digital switches operate on TDM signals).

7.11.1 Communications Satellite Networks Security

Geostationary satellite networks offer terrestrial coverage that excludes the polar regions, in contrast to LEOS and MEO networks that include the polar regions (90N/90S) as well as all ocean areas. These systems use short burst data with end-to-end encryption, signaling for sensitive user data, and denial of service protection.

However, the initial deployment of satellites as a viable and cost-efficient communications network in the sky has moved within the last decade from government and military to civilian applications, and with the increase of security breaches in other communications technologies led space agencies to develop security requirements for protocols, interfaces and data structures of satellite communication networks. In particular, security for all three major segments—ground, space and control—have been under scrutiny for security.

- The ground segment provides a secure reference architecture, for which global threats and vulnerabilities are identified. It also performs a risk assessment with possible solutions that are evaluated for transparency, implementation, performance, and for conformance to standards.

- The space segment is using a number of space-related protocols (see, for example, www.wiki.uni.lu/secan-lab) that are defined by the Consultative Committee for Space Data Systems or CCSDS (www.ccsds.org). CCSDS defines the space communication protocol standard (SCPS) that support security features by default. Note that CCSDS is a multinational forum that was founded by the world's major space agencies.

- The control segment forms the ground endpoint for all communication with a spacecraft and plays a central role in the ground segment infrastructure.

Based on the OSI model, security standards are defined for the application, transport, network, data link (protocol sublayer) and channel (coding sublayer), and the physical layer (RF and modulation) (Figure 7.16). Among these, SCPS is a suite

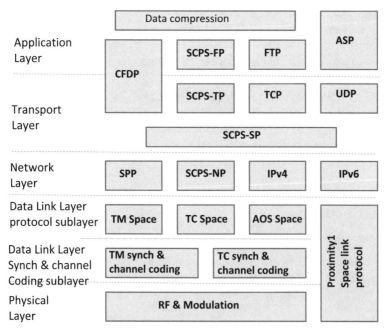

Figure 7.16.   The CCSDS protocol stack and related protocols.

of open protocol standards designed to meet the requirements of commercial, educational, and military environments. Originally they were jointly developed by NASA and DoD's USSPACECOM and today they exist as final Recommendations of CCSDS standards, United States DoD Military Standards, and ISO standards. The SCPS suite includes:

- The SCPS File Protocol (SCPS-FP) for uploading spacecrafts with commands and software, and downloading collections of telemetry data. The SCPS-FP is based on the Internet File Transfer Protocol (FTP) [48].

- The SCPS Transport Protocol (SCPS-TP) provides reliable end-to-end delivery of spacecraft command and telemetry messages between computers that are communicating over a network that may contain unreliable space data transmission paths. The SCPS-TP is based on the well-known Internet Transmission Control Protocol (TCP) [49].

- The SCPS Security Protocol (SCPS-SP) provides end-to-end security and integrity of message exchange. The SCPS-SP is derived from the Secure Data Network (SDNS) "SP3" protocol, the ISO Network Layer Security Protocol (NLSP), the Integrated Network Layer Security Protocol (I-NLSP), and the Internet Engineering Task Force's (IETF) Internet Protocol Security (IPSEC) Encapsulating Security Payload (ESP) and Authentication Header (AH) protocols [50].

- The SCPS Network Protocol (SCPS-NP) supports both connectionless and connection-oriented routing of messages through networks that have space or other wireless data links. The SCPS-NP is based on the well-known Internet Protocol (IP), with modifications to support new space routing needs and increased communications efficiency [51].

7.12 WIRELESS AD-HOC NETWORKS

A wireless ad hoc network is a topologically wireless mesh network in which neighboring nodes keep changing due to node movement. As such, the neighbors to a node are temporary and the determination of the next node (for routing purposes) is dynamic and depends on the relative movement of nodes with respect to a reference node. The interconnecting links between nodes are typically wireless, although under certain circumstances, free space optical and infrared is possible. The networks are termed *ad-hoc* because there is no priority difference among nodes in the network.

A wireless ad hoc network has no infrastructure and every node can forward data according to dynamically changing connectivity tables. Thus, ad-hoc networks communicate from node to node, and not like the wireless cellular via a base station; there is no base station in ad-hoc networks. A consequence is that a node communicates with another node in the ad-hoc network over multiple single hops (Figure 7.17). Thus, routing and control operations are executed by each node on the path.

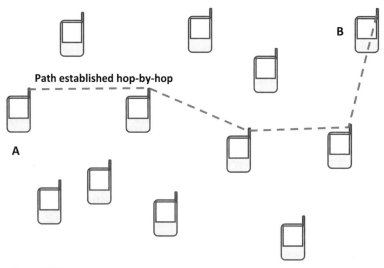

Figure 7.17.   A wireless ad hoc network consists of nodes that move with respect to each other and thus its topology changes with time. As a consequence, the minimal routes between two same nodes are not through the same nodes.

There are two major ad-hoc networks categories:

- *Mobile Ad Hoc Networks (MANET).* The nodes are mobile and they frequently break away or enter a network structure; this requires more resources as well as more complex auto-discovery, and routing protocols are required.

- *Wireless Sensor Networks (WSN).* WSN consists of nodes with limited resources, such as limited battery power and life, memory, computational power, and protocol complexity. WSN nodes do not necessarily move but they are often densely deployed in a specific geographical area to measure physical phenomena such as temperature, pressure, flux, speed, and so on. WSNs may also be deployed to detect movement and possible attacks.

Each node in the wireless ad-hoc network partakes in the route discovery when a node needs to communicate with another. Because of the node mobility, the non-stationary network composition (or, the network topology changes frequently and aperiodically), nodes need to know at any time their neighboring nodes and their characteristics so that they can determine the next best node in the multi-hop route. Therefore, to determine the best path or route across nodes, more sophisticated routing protocols are needed along with connectivity tables that are periodically updated. The proactive protocol and the reactive protocol are among the protocols proposed to date.

- The proactive protocol is table-driven, maintains routing tables for all nodes in the ad-hoc network, updates these tables frequently, and requires complex protocols and computational power. It is applicable to networks, the configuration and composition of which frequently change. Among the proactive protocols is the optimized link state routing protocol (OLSR) [52], the topology dissemination based on reverse-path forwarding (TBRPF) [53], and the secure efficient ad-hoc distance vector routing protocol (SEAD) [54].

- The reactive protocol maintains routing tables only for the nodes that are on the route to the destination node, and not of all nodes in the ad-hoc network. Clearly, this case is applicable to networks that either do not change or they change slowly, as is the case for stationary sensor networks, and route discovery is much simpler and much faster. Among the reactive protocols is the dynamic source routing protocol (DSR) [55], the ad-hoc, on-demand distance vector routing protocol (AODV) [56], the authenticated routing for ad-hoc networks (ARAN) [57]. In the latter protocol, a trusted certificate server assigns a certificate encrypted with the server's public key to each node in the network. The certificate contains the node IP address, the node public key, when the certificate was created, and when the certificate expires.

Wireless ad-hoc networks are flexible, self-organizing, and self-configurable, structureless, mobile, cost-efficient, and easy to deploy. Therefore, they find applications in the military (battlefield, submarine detection), disaster areas (searching units), collaborative operations (classroom, conferences), terrorist attack detection (multiple array of sensors), disaster detection (tsunami, seismic activity), health care to detect abnormalities, migration studies, and more.

7.12.1 Wireless Ad-Hoc Network Security

Wireless ad-hoc networks are challenged with similar security breaches as wireless cellular technology. In addition, because wireless ad-hoc nodes are small portable devices with limited resources, the network uses distributed control and multi-hop to transport traffic. End-to-end security is more vulnerable; a bad actor may exploit the weakest node or link on a route and attack the network. Thus, attacks are categorized as passive, active, external, and internal.

- In *passive attacks*, the attacker is near the network and copies information in an attempt to find the encryption keys, and also operation, administration, and maintenance (OAM) information that can be used to future attacks.
- In *active attacks*, the attacker attempts to capture, delete, alter, or redirect a message. The attacker can also cause denial of service and resource exhaustion.
- In *external attacks*, the attacker is external to the group of nodes that comprises the ad-hoc network. An external attacker may impersonate or mimic a node of the network and cause, among other problems data rerouting so that data does not reach its designated destination.

- In *internal attacks*, an attacker is from within the logical communicating group of nodes. In this case, it is more difficult to detect the security breach since all nodes of the network are equivalent. In this category, one special type of attack is the *node misbehavior* and another *node selfishness*. Node misbehavior is defined as the unauthorized behavior of an internal node that does not provide the expected cooperation by its group nodes. Node misbehavior may be the result of node malfunctioning or corrupted control software [58] and it impacts the MAC layer causing routing performance. Node selfishness may be the result of a malicious user in an attempt to gain access or disrupt service; forces a node to cheat or abuse accessing and routing rules in order to gain preferential access to resources, such as bandwidth and priority.

Additionally, wireless sensor network may be under a number of possible attacks. For example:

- A sensor node can be captured by attackers and data stored in it can be read. Stored information may be private messages, encryption keys, and control information.
- A sensor node may be disabled or malfunction when under attack.
- A fake node with fake ID may be injected into the network and pretend it is a valid node. This is serious form of attack because the fake node can perform many types of malicious acts without being suspected.

Authentication and key management are as important in ad-hoc networks as in any other network. If an attacker can get the trust of a node and the encryption keys, then all information through the node is visible to the attacker, including the key distribution and the key renewal. In ad-hoc networks, authentication and key management is challenging because nodes may dynamically be relocated changing the neighborhood landscape; that is, neighboring nodes at one time may not be the same after a while. To envision this, think of a group of honey bees as nodes of an ad-hoc network in a flower garden. With reference to a particular honey-bee, its neighboring bees keep changing continuously as they search for flowers with the most nectar. When a bee discovers a flower patch with plenty of nectar, it is able to convey its coded messages to the next bee, avoiding other insects that do not belong to the group. That is, each bee is able to authenticate (or recognize) the neighboring bees and send its coded message, which in turn do the same with their own neighbors, and so on.

In the ad-hoc network all nodes are the same and there are no routers, switches, or gateways with special functions as in other networks. That is, structurally the ad-hoc network is not hierarchical or three-dimensional, but horizontal like a flat plane. Thus, there is no particular node that functions like a trusted third party to act like a certification authority, or a key distribution center, or an arbiter, and therefore, several challenges need to be overcome. For example, if symmetric cryptography is used, there is no certification authority to set up secret keys, but instead assign every pair of nodes during the initial configuration a pair of secret keys. This, however, is a tedious process if the network consists of many nodes. Similarly, if public key (asymmetric) cryptography is used, key distribution and key renewal become very difficult because these processes rely on a node with special functionality. Wireless sensor ad-hoc networks have the following characteristics:

- The network can have a central infrastructure, such as a base station, that serves as a gateway to a traditional network.
- The number of nodes in the network is several orders of magnitude higher than that in a mobile ad hoc network.
- Wireless sensor-nodes are usually densely deployed in a geographical area.
- Wireless sensor-nodes are used mainly for measurements (telemetry data and thus low data rate), whereas nodes in mobile ad hoc networks are used for data communication (high data rate).
- Wireless sensor-nodes have limited resources and thus lower computational intelligence and shorter transmission range, than nodes in an ad-hoc mobile network.
- Wireless sensor-nodes are not necessarily mobile.
- Each wireless sensor-node in the network does not have a global identification (IP address), although the network itself may have one (each node in mobile ad hoc network has an IP address).

The wireless ad-hoc sensor networks are typically deployed in mission-critical situations. They are vulnerable to attacks and therefore security is of importance. However, sensor-nodes have limited resources. Thus computational intensive public key cryptography, symmetric cryptography, authentication, and digital signatures are prohibitive. Despite this, wireless ad-hoc sensor networks should be armored with more security requirements than other networks. In addition to traditional security features, they should be equipped with forward secrecy and with backward secrecy. That is, when a node departs from the network, it should not be able to receive or transmit information in the same network; this is forward secrecy. Similarly, a newly added node should not be able to receive information from or transmit it to the network prior to joining it, and it should not be able to communicate with the network it came from after joining the new network; this is backward secrecy.

Key distribution in wireless sensor networks is important because without it, encryption, decryption, and authentication cannot be performed. Among the proposed key distribution methods, some are:

- The network-wise key distributes a single key to all the nodes of the network during the initial setup. Because the security of the entire network depends on a single key, security is vulnerable and the method is suitable to networks with low communication security requirements.
- The pair-wise key method assigns a pair-wise key to each pair of nodes during the initial setup. That is, for a network with N nodes, each node needs to store

($N - 1$) pair-wise keys, wasting resources because only neighboring nodes communicate. In addition, this method becomes problematic when new nodes join or depart the network.

- The centralized key management makes use of a base station as the key distribution center. Public key cryptography and symmetric cryptography work well in this case. However, a base station is contrary to ad-hoc network philosophy, which assumes that all nodes are the same.

- The distributed key management assumes a number of nodes to be involved in a cooperative key distribution. Distributed key management is more scalable than centralized key management. As the number of the nodes in the network increases, so do the number of nodes involved in key distribution. However, the idea of key distributing nodes and common nodes is contrary to the ad-hoc network philosophy, which assumes that all nodes are the same.

Although certain standards have been issued [59–62], research in the field of ad-hoc and sensor networks continues [63–81]. New protocols and new cryptographic methods are being proposed. Ad-hoc networks are attractive because of their topological flexibility, but traffic routing and traffic balancing, routing optimization, data privacy, node redundancy and network security are important issues that need to be carefully investigated and solidified.

REFERENCES

1. T. KARYGIANNIS and L. OWENS, "*Wireless Network Security*," National Institute of Standards and Technology (NIST), Nov. 2002, pp. 4.1–4.13.
2. IEEE 802.11, 1999 Edition (ISO/IEC 8802-11: 1999) IEEE Standards for Information Technology—Telecommunications and Information Exchange between Systems—Local and Metropolitan Area Network—Specific Requirements—Part 11: Wireless LAN Medium Access Control (MAC) and Physical Layer (PHY) Specifications.
3. IEEE 802.11a-1999 (8802-11:1999/Amd 1:2000(E)), IEEE Standard for Information technology—Telecommunications and information exchange between systems—Local and metropolitan area networks—Specific requirements—Part 11: Wireless LAN Medium Access Control (MAC) and Physical Layer (PHY) specifications—Amendment 1: High-speed Physical Layer in the 5 GHz band.
4. IEEE 802.11b-1999, Supplement to 802.11-1999, Wireless LAN MAC and PHY specifications: Higher speed Physical Layer (PHY) extension in the 2.4 GHz band.
5. 802.11b-1999/Cor1-2001, IEEE Standard for Information technology—Telecommunications and information exchange between systems—Local and metropolitan area networks—Specific requirements—Part 11: Wireless LAN Medium Access Control (MAC) and Physical Layer (PHY) specifications—Amendment 2: Higher-speed Physical Layer (PHY) extension in the 2.4 GHz band—Corrigendum1.
6. IEEE 802.11d-2001, Amendment to IEEE 802.11-1999, (ISO/IEC 8802-11) Information technology–Telecommunications and information exchange between systems–Local and metropolitan area networks–Specific requirements–Part 11: Wireless LAN Medium Access Control (MAC) and Physical Layer (PHY) Specifications: Specification for Operation in Additional Regulartory Domains.

7. IEEE 802.11e-2005, IEEE Standard for Information technology—Telecommunications and information exchange between systems—Local and metropolitan area networks—Specific requirements Part 11: Wireless LAN Medium Access Control (MAC) and Physical Layer (PHY) specifications: Amendment 8: Medium Access Control (MAC) Quality of Service Enhancements.

8. IEEE 802.11g-2003, IEEE Standard for Information technology—Telecommunications and information exchange between systems—Local and metropolitan area networks—Specific requirements—Part 11: Wireless LAN Medium Access Control (MAC) and Physical Layer (PHY) specifications—Amendment 4: Further Higher-Speed Physical Layer Extension in the 2.4 GHz Band.

9. IEEE 802.11h-2003, IEEE Standard for Information technology—Telecommunications and Information Exchange Between Systems—LAN/MAN Specific Requirements—Part 11: Wireless LAN Medium Access Control (MAC) and Physical Layer (PHY) Specifications: Spectrum and Transmit Power Management Extensions in the 5GHz band in Europe.

10. IEEE 802.11i-2004, Amendment to IEEE Std 802.11, 1999 Edition (Reaff 2003). IEEE Standard for Information technology–Telecommunications and information exchange between system–Local and metropolitan area networks, Specific requirements–Part 11: Wireless LAN Medium Access Control (MAC) and Physical Layer (PHY) specifications–Amendment 6: Medium Access Control (MAC) Security Enhancements.

11. IEEE 802.11j-2004, IEEE Standard for Information technology—Telecommunications and information exchange between systems–Local and metropolitan area networks—Specific requirements—Part 11: Wireless LAN Medium Access Control (MAC) and Physical Layer (PHY) specifications—Amendment 7: 4.9 GHz–5 GHz Operation in Japan.

12. IEEE 802.11 Working Group (2007-06-12), *IEEE 802.11-2007: Wireless LAN Medium Access Control (MAC) and Physical Layer (PHY) Specifications,* ISBN 0-7381-5656-9.

13. C. PERKINS and R. BHAGWAT, "Highly Dynamic Destination-sequenced Distance-vector Routing (DSDV) for Mobile Computer," in *Proceedings of the Conference on Communications Architectures, Protocols and Applications*, New York, NY, 1994, pp. 234–244.

14. WI-FI ALLIANCE. (2003), "Wi-Fi Protected Access: Strong, standards-based, interoperable security for today's Wi-Fi networks", retrieved March 1, 2004,

15. IETF RFC 3748: Extensible Authentication Protocol (EAP), June 2004, http://tools.ietf.org/html/rfc3748.

16. Amendment to IEEE 802.11, 1999 Edition (Reaff 2003). "IEEE Standard for Information technology–Telecommunications and information exchange between system–Local and metropolitan area networks—Specific requirements—Part 11: Wireless LAN Medium Access Control (MAC) and Physical Layer (PHY) specifications–Amendment 6: Medium Access Control (MAC) Security Enhancements", IEEE 802.11i-2004.

17. IETF RFC 2865: Remote Authentication Dial In User Service (RADIUS), June 2000. http://tools.ietf.org/html/rfc2865.

18. "Two Billion GSM Customers Worldwide", http://www.prnewswire.com/cgibin/stories.pl?ACCT=109&STORY=/www/story/06-13-2006/0004379206&EDATE.

19. S.V. KARTALOPOULOS, "A Primer on Cryptography in Communications", *IEEE Communications Magazine*, vol. 44, no. 4, 2006, pp. 146–151.

20. "CDMA2000 Operators Delivered 3G Services to More Than 350 Million Subscribers Worldwide in Q1 2007", http://www.primenewswire.com/newsroom/news.html?d=120585.

21. R. STEELE, C.-C. LEE, and P. GOULD, *GSM, cdmaOne and 3G Systems*, 1st ed., Wiley, 2001.

22. J. EBERSPAECHER, H.-J. VOEGEL, and C. BETTSTETTER, *GSM Switching, Services and Protocols*, 2nd ed., Wiley, 2001.

23. 3GPP TS 55.216 v.6.2.0, 9/2003, "3rd Generation Partnership Project; Technical Specification Group Services and System Aspects; 3G Security; Specifications of the A5/3 Encryption Algorithms for GSM and ECSD, and the GEA3 Encryption Algorithm for GPRS; Document 1: A5/3 and GEA3 Specifications, Release 6.

24. www.gsm-security.net/gsm-security-faq.shtml.

25. A.R. PRASAD, A. ZUGENMAIER, and P. SCHOO, "Next generation communications and secure seamless handover", *Security and Privacy for Emerging Areas in Communication Networks*, 2005. Workshop of the 1st International Conference, September 5–9, 2005, pp. 267–274.

26. M. STEER, "Beyond 3G", *IEEE Microwave Magazine*, vol. 8, no. 1, February 2007, pp. 76–82.

27. K. MUROTA, "Mobile Communications Trends in Japan and DoCoMo's Activities Towards 21st Century," Proc. AMOS, June 1999.

28. J. QADDOUR and R.A.C. BARBOUR, "Evolution to 4G wireless: problems, solutions, and challenges", *Computer Systems and Applications*, 2005. The 3rd ACS/IEEE International Conference, 2005, p. 78.

29. 3GPP Technical Specifications, 3G Security; Wireless Local Area Network (WLAN) interworking security, TS33.234 v6.1.0, R6, Third Generation Partnership Project, Jun. 2004.

30. A.R. PRASAD and P. SCHOO, "IP Security for Beyond 3G Towards 4G", http://www.docomoeurolabs.de/pdf/publications/STL-ipsec_WWRF7_02.pdf.

31. "An insight into IEEE 802.16 WiMax", retrieved January 30, 2008, from www.seas.gwu.edu/~cheng/388/LecNotes2007/80216WiMAXSecurity.pdf.

32. "802.16 Security", retrieved January 30, 2008, from www.cs.tut.fi/~83180/83180_05_S10c.ppt

33. IEEE 802.16TM-2001, IEEE Standards for Local and Metropolitan Area Networks, IEEE, 2002.

34. R. HOUSLEY et al. "Internet X.509 Public Key Infrastructure Certificate and Certificate Revocation List (CRL) Profile", RFC 3280, IETF (Internet Engineering Task Force), April, 2002.

35. *Introduction to IMS*, Ericsson White Paper, March 2007.

36. *Motorola IP Multimedia Subsystem*, Motorola White Paper, November 2005.

37. M. POIKSELKA, G. MAYER, H. KHARTABIL, and A. NIEMI, "*The IMS: IP Multimedia Concepts and Services in the Mobile Domain*", Wiley Publishers, 2004.

38. G. CAMARILLO and M.A. GARCIA-MARTIN, "*The 3G IP Multimedia Subsystem (IMS): Merging the Internet and the Cellular worlds*", Wiley Publishers, 2004.

39. S.I.G. BLUETOOTH, "How *Bluetooth* Technology Works", retrieved January 31, 2008, http://www.bluetooth.com/Bluetooth/Technology/Works/.

40. T. ZIMMERMAN, "Personal Area Networks: Near-field intrabody communication", MIT Media Lab., vol. 35, no. 3&4, 1996.

41. O. SHIVERS, "BodyTalk and the BodyNet: A Personal Information Infrastructure", Personal Information Architecture Note 1, MIT Laboratory for Computer Science, Cambridge, MA, December 1, 1993.

42. M. WEISER, "The Computer for the 21st Century," Scientific American, vol. 265, No. 3, 94–104, September 1991.

43. E.R. POST, M. REYNOLDS, M. GRAY, J. PARADISO, and N. GERSHENFELD, "Intrabody buses for data and power", Physics and Media, MIT media Lab.

44. S.V. KARTALOPOULOS, "A Global Multi-satellite Network", Proceeding of the International Communications Conference (ICC'97), Montreal, Canada, 1997, pp. 699–698.

45. S.V. KARTALOPOULOS, "A Global Multi Satellite Network", US patent 2/11/1997, 5,602,838

46. G. MARAL and M. BOUSQUET, *Satellite Communications Systems*. New York, John Wiley and Sons, 2002.

47. K.S. CHOI, J.H. JO, C.S. SHIN, and S.P. LEE, "Conceptual Design of the Network Control Functions for OBS Satellite Communication System," in *Advanced Communication Technology*, 2005, pp. 641–645.

48. "*SCPS File Protocol Overview*": CCSDS Standard: 717.0-B-1, DoD MIL Standard: MIL-STD-2045-47000, ISO Standard: ISO 15894; also in www.scps.org/Documents/.

49. "*SCPS Transport Protocol Overview*": CCSDS Standard: 714.0-B-2; DoD MIL Standard: MIL-STD-2045-44000; ISO Standard: ISO 15893; also in www.scps.org/Documents/.

50. "*SCPS Security Protocol Overview*": CCSDS Standard: 713.5-B-1; DoD MIL: MIL-STD-2045-43001; ISO Standard: ISO 15892; also in www.scps.org/Documents/.

51. "*SCPS Network Protocol Overview*": CCSDS Standard: 713.0-B-1; DoD MIL: MIL-STD-2045-43000; ISO Standard: ISO 15892; also in www.scps.org/Documents/.

52. IETF RFC 3626, http://www.ietf.org/rfc/rfc3626.txt?number=3626.

53. IETF RFC 3684, http://www.ietf.org/rfc/rfc3684.txt?number=3684.

54. Y.-C. HU, D. JOHNSON, and A. PERRIG, "SEAD: Secure Efficient Distance Vector Routing for Mobile Wireless Ad Hoc Networks," *Ad Hoc Networks, Elsevier*, vol. 1, no. 1, pp. 175–92, 2003.

55. IETF RFC 4728, http://www.ietf.org/rfc/rfc4728.txt?number=4728.

56. IETF RFC 3561, http://www.ietf.org/rfc/rfc3561.txt?number=3561.

57. K. SANZGIRI et al., "A Secure Routing Protocol for Ad Hoc Networks," *Proceedings of the 10th IEEE International Conference on Network Protocols*, Washington, DC, Nov. 2002, pp. 78–89.

58. L. GUANG, C. ASSI, and A. BENSLIMANE, "On MAC Layer Misbehaving in Wireless Networks: Challenges and Solutions", IEEE Wireless Communications Magazine, vol. 15, 2008, to be published.

59. IETF RFC 4728, http://www.ietf.org/rfc/rfc4728.txt?number=4728.

60. IETF RFC 3561, http://www.ietf.org/rfc/rfc3561.txt?number=3561.

61. IETF RFC 3626, http://www.ietf.org/rfc/rfc3626.txt?number=3626.

62. IETF RFC 3684, http://www.ietf.org/rfc/rfc3684.txt?number=3684.

63. C.-K. TOH, "A Novel Distributed Routing Protocol to Support Ad Hoc Mobile Computing", *Proceedings of IEEE 15th Annual International Phoenix Conference on Computers and Communications, Mar. 1996*, pp. 480–486.

64. B. DAVID and A. DAVID, "Dynamic Source Routing in Ad Hoc Wireless Networks", *Mobile Computing*, Kluwer Academic, 1996.

65. S. MURTHY and J.J. GARCIA-LUNA-ACEVES, "An Efficient Routing Protocol for Wireless Networks", *ACM Mobile Networks and Application Journal*, Special Issue on *Routing in Mobile Communications Networks*, 1996, pp. 183–197.

66. V.D. PARK and M.S. CORSON, "A Highly Adaptive Distributed Routing Algorithm for Mobile Wireless Networks", *Proceedings of INFOCOM'97*, Apr. 1997, pp. 1405–1413.

67. S. YI, R. NALDURG, and R. KRAVETS, "Security-aware Ad-hoc Routing for Wireless Networks". *Proceedings of the 2nd ACM International Symposium on Mobile Ad Hoc Networking & Computing*, New York, NY, 2001, pp. 299–302.

68. K. SANZGIRI et al., "A Secure Routing Protocol for Ad Hoc Networks", *Proceedings of the 10th IEEE International Conference on Network Protocols*, Washington, DC, Nov. 2002, pp. 78–89.

69. P. PAPADIMITRATOS and Z. HAAS, "Secure Routing for Mobile Ad Hoc Networks", *SCS Communication Networks and Distributed Systems Modeling and Simulation Conference*, San Antonio, Texas, Jan. 2002.

70. C. SIVA RAM MURTHY and B.S. MANOJ, *Ad Hoc Wireless Networks—Architectures and Protocols*, Prentice Hall, NJ, USA, 2004.

71. W. STALLINGS, *Cryptography and Network Security Principles and Practices*, 4th ed., Pearson Education Inc., NJ, USA, 2006.

72. C. KARLOF and D. WAGNER, "Secure Routing in Wireless Sensor Networks: Attacks and Countermeasures", *Elsevier's Ad Hoc Network Journal, Special Issue on Sensor Network applications and Protocols*, vol. 1, pp. 293–315, September 2003.

73. S. CAPKUN, L. BUTTYAN, and J.P. HUBAUX, "Self-organized Public-key Management for Mobile Ad Hoc Networks", *IEEE Transactions on Mobile Computing*, vol. 2, no. 1, pp. 52–64, 2003.

74. D. DJENOURI and L. KHELLADI, "A Survey of Security Issues in Mobile Ad Hoc and Sensor Networks", *IEEE Communications Survey & Tutorials*, vol. 7, no. 4, pp. 2–28, 2005.

75. Y.H.W. LEE, "A Cooperative Intrusion Detection System for Ad Hoc Networks", *Proceedings of the 1st ACM Workshop on Security of Ad Hoc and Sensor networks*, Fairfax, Virginia, USA, 2003, pp. 135–147.

76. Y. ZHANG, W. LEE, and Y. HUANG, "Intrusion Detection Techniques for Mobile Wireless Networks", *ACM Wireless Networks*, vol. 9, no. 5, pp. 545–556, September 2003.

77. O. KACHIRSKI and R. GUHA, "Effective Intrusion Detection Using Multiple Sensors in Wireless Ad Hoc Networks", *Proceedings of the 36th Annual Hawaii International Conference on System Sciences*, Big Island, Hawaii, January 2003, pp. 57.1.

78. Y. XIAO (Ed), *Security in Distributed, Grid, and Pervasive Computing*, Auerbach Publications, CRC Press, 2006.

79. J.N. Al-Karaki and A.E. Kamal, "Routing Techniques in Wireless Sensor Networks: A Survey", *IEEE Wireless Communications*, vol. 11, no. 6, pp. 2–28, December 2004.

80. L. Eschenauer and V.D. Gligor, "A Key-management Scheme for Distributed Sensor Networks", *Proceedings of the 9th ACM Conference on Computer and Communications Security*, New York, NY, 2002, pp. 41–47.

81. Y. Sang et al., "Secure Data Aggregation in Wireless Sensor Networks: A Survey", *Proceedings of the Seventh International Conference on Parallel and Distributed Computing, Applications and Technologies*, Washington, DC, 2006, pp. 315–320.

Chapter 8

Network Security: Wired Systems

8.1 INTRODUCTION

The previous Chapter examined the most popular wireless communication technologies, systems, and protocols, and security aspects in terms of source identification, validation, network access authorization, and encryption keys.

The traditional wired telecommunication network, also known as the public switched telecommunications network (PSTN), was designed several decades ago with copper pair wires as the transmission medium, at both the loop and the trunk plants. This network was retrofitted or updated using new technology to cope with increasing traffic demand and new technological developments, particularly the circuit transistor integration to single chips (IC), digital transmission and switching, and computers as control units. In contrast to the Internet network, computers in the PSTN nodes were isolated from customer data traffic, and therefore security breaches and network attacks, as encountered in the Internet, were absent. The exception was eavesdropping and illegally placing long distance phone calls "for free," or by placing charges on another subscriber.

The PSTN controller is isolated from customer data; data is expressly passed from the input node to an output via a switch according to a pre-setup connection. In contrast, in the Internet network, packetized information is stored in the computer memory from where it is passed to an output according to a routing protocol. Additionally, customer traffic paths and control message paths (for node configuration and node control, known as Operations, Administration and Management (OA&M)) were not the same as they are with the Internet); in fact, control messages are assembled according to standardized telecommunications control protocols and control messages enter their own physical port, which is different than that of customer traffic. OA&M messages are transmitted locally or remotely over the Intranet with authentication mechanisms such as source identification, passwords, etc., so that it is not easy for an intruder to gain access of the control path of nodes in the network. Thus, it is the opinion of many that the Internet security weakness is

Security of Information and Communication Networks, by Stamatios V. Kartalopoulos
Copyright © 2009 Institute of Electrical and Electronics Engineers

because of the common paths for customer data and control messages, and also that Internet nodes are not switching nodes (although the latter is addressed with modern switch-based Internet nodes that execute label switching protocols).

In the 1980s, wire-based trunks of the PSTN network started been replaced with fiber-optic cables and a new standardized protocol was developed, based on which a new breed of nodes and network was designed; in Europe, the protocol and nodes were known as synchronous digital hierarchy (SDH) and in the USA as synchronous optical network (SONET) [1, 2]. Thus, transmission was now based on photons and not electrons, and because of the THz frequency of photons, optical signals with wavelength of 1310 nm and also 1550 nm could be modulated to carry many times more than 43 Mbps, which the wire carried (Table 8.1). That is, optical signals were now able to exceed the 1 Gbps mark, a data rate that was at the time very challenging. However, for compatibility and other technical reasons (such as synchronization), specific rates were defined, known as SDH or SONET rates (Table 8.2).

SDH/SONET was proved so efficient and cost-effective for synchronous payloads that it rapidly replaced all traditional wired-based networks; to date, the fiber-optic medium is the only one deployed on the trunk side of the network in all advanced countries, including transoceanic (submarine).

Because of the huge amount that single optical channel can carry, and with the development of optical and photonic technology, particularly lasers, photodetectors, and optical amplifiers, many optical channels, each at different wavelength, were multiplexed and coupled onto a single fiber, and the wavelength division multiplex (WDM) technology was applied to a network, so that now a single fiber carries more than 1 Tbps [3–5]. As a result of the huge amount of information that could be

Table 8.1. Traditional bitrates of the wired synchronous hierarchy

Facility	U.S.	Europe	Japan
DS0/E0	64 kbps	64 kbps	64 kbps
DS1	1,544 kbps		1,544 kbps
E1		2,048 kbps	
DS1c	3,152 kbps		3,152 kbps
DS2	6,312 kbps		6,312 kbps
E22		8,448 kbps	
			32,064 kbps
E31		34,368 kbps	
DS3	44,736 kbps		
DS3c	91,053 kbps		
			97,728 kbps
E4		139,264 kbps	
DS4	274,176 kbps		
			397.2 Mbps

Table 8.2. Bitrates of the synchronous fiber-optic network

Signal Designation			
SONET	SDH	Optical	Line Rate (Mbps)
STS-1	**STM-0**	**OC-1**	**51.84 (52 M)**
STS-3	**STM-1**	**OC-3**	**155.52 (155 M)**
STS-12	**STM-4**	**OC-12**	**622.08 (622 M)**
STS-48	**STM-16**	**OC-48**	**2,488.32 (2.5 G)**
STS-192	**STM-64**	**OC-192**	**9,953.28 (10 G)**
STS-768	**STM-256**	**OC-768**	**39,813.12 (40 G)**

OC-N: Optical Carrier-level N
STS-N: Synchronous Transport Signal-level N
STM-N: Synchronous Transport Module-level N

transported over the fiber-optic backbone network, a problem emerged known as access "bottleneck," or the first/last mile problem. That is, although fiber trunks could potentially carry a huge amount of data, that amount could not be delivered over the wired-based loops. To alleviate this, a digital transmission technology was developed that was able to surpass the 64 Kbps limitation on the traditional twisted pair loop to more than Mbps; this technology was called digital subscriber line (DSL), which depending on the method was subdivided to ADSL (A for asymmetric), VDSL (V for very high speed), and so on.

WDM technology, along with the newly developed water-free fiber, makes it possible to deploy the fiber medium in the loop plant replacing the twisted copper pair and to deliver data to the home or small business at a rate that ranges from several Mbps or more than 1 Gbps.

With regard to security, the SDH/SONET kept the same philosophy of the traditional network; switching nodes and customer data and OA&M messages over separate physical paths. In addition, because the SDH/SONET payload is time division multiplexed according to a complex byte multiplexing scheme, isolating a particular channel in order to eavesdrop requires specialized know-how and specialized equipment; this, however, does not mean that tapping a fiber cable and eavesdropping is impossible. It is just more difficult.

Although wired networks are not used, many attributes are either still in use or they set the foundation of optical networks. Therefore, this chapter examines wire networks and their security, and also optical networks and their security.

8.2 WIRED NETWORKS

The wired network consists of loops that connect end devices (telephones) with switching nodes, known as exchanges. Nodes are connected with other nodes in a mesh topology with trunks (also known as junctions) to form what is known as the backbone network.

8.2.1 The Loop Plant and the Trunk Plant

A loop consists of a twisted pair (TP) of copper wires insulated with paper. Typical loop lengths vary from few hundred meters to more than 6 Km, that is, the distance between telephone and exchange.

The wire gauge of loops may not be the same; depending on loop length, it has different diameter as it is related with Ohmic resistance per 1000 feet (or 333 meters) by the relationship:

$$R_{dc}(\text{W}) = 0.1095/d^2$$

where R_{dc} is the *dc* resistance and d the wire diameter.

In the U.S., wire diameter is expressed in gauge (ga.). Popular wire gauges used in loops are shown in Table 8.3.

As an example, 26 ga. wire has 0.405 mm dia, a resistance of 83.5 Ohms and 0.51 dB loss per 1000 feet of length. Thus, often loops are referred to by their Ohmic resistance. For example:

- If the wire gauge is 26, then a loop of 2 dB loss implies a 4000 feet loop.

- If the gauge is not known, then a "83.5 Ohms loop" implies 26 gauge

Typically, the loop loss is limited to 6 dB. Then, the maximum loop length limit is determined based on wire gauge (Table 8.4).

The trunk plant consisted of two wires interconnecting network nodes. However, the requirements for trunks were different from loops.

Table 8.3. Popular gauges and corresponding wire diameter

Gauge (ga.)	Wire diameter (mm)
26	0.405
24	0.511
22	0.644
19	0.910

Table 8.4. Typical wire gauges at the loop plant, each gauge meeting specific loop lengths

Gauge (ga.)	Loop length (Km)
26	3.9
24	4.85
22	6.33
19	9.5

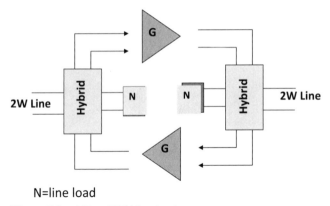

N=line load

Figure 8.1.   2 W to 4 W bidirectional repeaters.

Trunks transported the multiplexed information of many users and thus higher data rates than the loops, such as 1.544 Mbps (known as T1 trunks) or higher and typically up to 44.736 Mbps (known as T3 trunks). Because of attenuation, noise and jitter, the signals on the trunks needed required every 2 Km repeaters to amplify, retime and reshape the signal. However, because of the two-way flow of information over two wires, the two directions had to be converted from two to four wires (two per directions) at the repeater site (Figure 8.1), using what is known as a hybrid circuit (which is based on transformers).

Having repeaters every 2 Km is both capital and maintenance intensive. With the increased demand for higher data rates, the wire trunks were quickly replaced by optical fiber; optical fiber delivers 1000 more data rate and requires an optical amplifier every 60–80 Km, and the fiber cable is much thinner than the wire cable. Thus, the fiber optic proved to be more cost-efficient in both capital expense (Capex) and operating expense (Opex).

8.2.2 Analog Loops and Digital Loops

The loop plant was initially engineered to transport analog signals that were band-limited between 300-to-3400 Hz (Figure 8.2); that is, loops were engineered to transport analog electrical signals that were the result of conversion spoken voice (acoustic pressure to electrical current).

The electrical signal of spoken voice (in contrast to singing voice that has an upper frequency limit above 10 KHz) was converted to digital at the network side of the loop by coder–decoders (CODEC) following a pulse coded modulation (PCM) weighted method known as μ-law (in the U.S.) or α-law (in Europe); the outcome of digitizing voice resulted in a 64 Kbps digital PCM signal, known as digital signal level-0 (DS0) (Figure 8.3). Thus, the digital format of analog signals started at the edge of the network (at the local exchange) where the CODECs were. From that entry point, several PCM signals were time division multiplexed (TDM) following

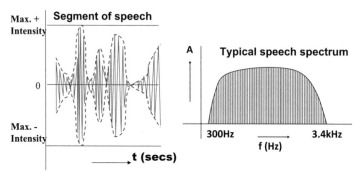

Figure 8.2. Spoken voice and band-limited spectrum.

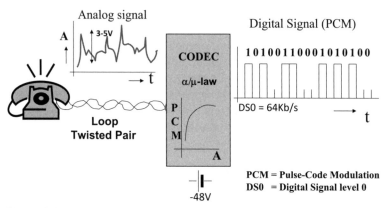

Figure 8.3. The analog signal from a telephone is converted to digital PCM by a CODEC at the entry point of the network.

a multiplexing hierarchy; in the U.S., the digital service 1 (DS1 at 1.544 Mbps) signal that consists of twenty-four 64 Kbps signals plus 8 Kbps for framing, data link and error control, whereas in Europe the E1 (at 2.048 Mbps) signal that consists of thirty 64 Kbps signals, plus 128 Kbps for framing, data link, and error control. These signals are further multiplexed resulting to higher data rates within the hierarchical public switched digital network (PSDN). The telecommunications hierarchy distinguishes two major types of switching nodes: one transports aggregate bandwidth between two points or cities over trunks—there are four switch classes, 1 to 4 (Figure 8.4)—and the other interfaces with the various end-devices (telephones) with loops (class 5 switch). Class 5 switch is the local exchange and it is also known as edge or as access node (Figure 8.5).

With the proven success of the optical network and with the spread of data networks in the 1980s, it was realized that the loop as engineered cannot meet the customer needs to transmit voice and data over the loop. As a consequence, a digital technology was developed [6–8] that was able to transmit digital signals over the

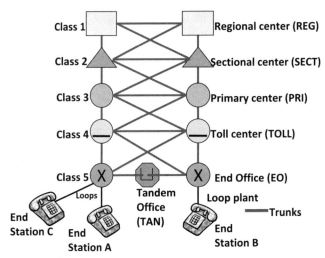

Figure 8.4. The switching hierarchy in USA; switching centers, classes and trunks. TAN denotes a Tandem switch that connects access nodes within the same serving area.

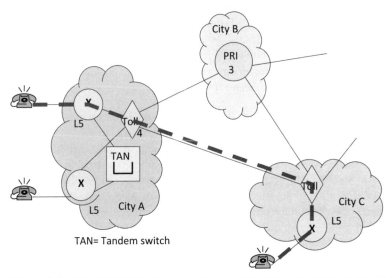

Figure 8.5. The PSDN distinguishes two fundamentally different switching nodes, the level 5 access and the higher level switches (toll, primary, etc). Connectivity is established across networks and from network to network.

loop, known as integrated services digital network (ISDN) [9–11]. With ISDN end devices, the end user is able to transmit 2×64 Kbps + 16 Kbps (or 144 Kbps), known as a basic rate interface (BRI) signal; however, ISDN BRI uses four wires or two twisted pairs. Thus, with ISDN and over the loop, the end user is able to transmit digital voice and data, or high speed data only and up to 144 Kbps; clearly, the

CODEC has moved in the ISDN telephone and the local exchange is retrofitted to recognize the ISDN signal and its comprehensive protocols.

However, although ISDN paved the way to digital loops, 20 years later the total ISDN BRI data rate was not enough; the end user now needs more bandwidth than 144 Kbps for voice and for very high speed data as a result of the exploding Internet and emerging data services. Thus, the digital subscriber loop (DSL) technology (in development since the 1990s) emerged as a high data rate over the loop [12–18].

In fact, DSL is a generic acronym with several family members, each transmitting at different data rates over different loop lengths and different modulation methods. Three of them are:

- *HDSL* or high bit-rate DSL. Using a single pair of wires, the downstream bit-rate is 768 Kbps, and the upstream bit-rate is 768 Kbps. The maximum length of loop is 4000 M.

- *ADSL* or asymmetric bit-rate DSL. The downstream bit-rate is much higher than the upstream. Using a single pair of wires, the downstream bit-rate is 1.5 Mbps, and the upstream is 176 Kbps. The maximum length of loop is 4000 M.

- *MSDSL* or multi-rate symmetric DSL. MSDSL-based systems are based on single pair SDL technology and offer one of many rates, and thus one of many loop lengths. For example, MSDSL on a 24 gauge unloaded copper pair provides service at 64/128 Kbps up to 8.9 Km, or 2 Mbps up to 4.5 Km. Unloaded copper wires means that inductors previously placed on analog loops (to suppress high frequencies) have been removed.

In fact, a typical DSL signal on the loop supports both analog voice and very high data rate, often called voice-over data (VoD) or ADSL over POTS (ADSLoPOTS). At the PSDN network side, Internet data from an ISP's network enters a DSL modem or DSL bridge. The output of the modem is capable of placing voice over data, so that both voice and data are supported. The combined signal (data and voice if a call is in progress) is coupled onto the loop. Thus, the signal on the loop is voice plus data, voice only or data only. To further clarify, the DSL path from the network to CPE may remain continuously on without affecting telephone service; when a call is coming in while the DSL data path is on, the telephone will ring and when the call is terminated, the DSL data service is still on; that is, the DSL bridge seamlessly connects two networks. Moreover, DSL provides shared Internet service; that is, more than one computer may be connected over the same DSL data path.

When used in residential or SOHO applications, the signal from the network is a combined analog (for voice) and digital (high bit rate). This is the signal that arrives at the wall connector. This signal is split in two paths, one path is connected to the DSL customer premises equipment (CPE), or the DSL modem, which provides Internet connectivity (via the RJ45 jack) to the computer, and the other path is connected to a low filter that passes the analog low frequency signal to the telephone.

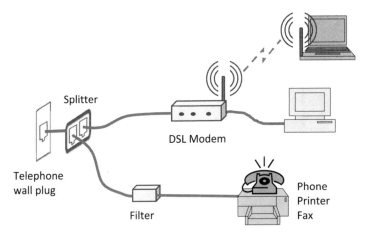

Figure 8.6. Connectivity of the DSL modem on the premises.

It is this combination of analog and digital signal that characterizes the DSL technology (Figure 8.6). An additional feature of the DSL modem is that, besides wired connectivity to a computer, it provides WiFi wireless connectivity.

From the network side of the DSL, there are two networks: one for synchronous (voice) traffic coming from a switch and another for data, coming from the Internet network. Traffic from the two networks merges on a device called DSL access multiplexer (DSLAM) where they are multiplexed. The DSLAM may be col-located with the switch at the central office, or be remotely located in an outside cabinet and behind the cabinet that provides connectivity to loops, known as main distribution frame (MDF); the latter case is known as serving area interface (SAI).

In the downstream direction (towards the customer), the DSLAM multiplexes voice and digital data and via the MDF they reach the DSL modem at the other end of the loop where the two signals are separated. In the upstream direction, digital data signals and low frequency voice signals propagate on the loop and via the MDF they reach the DSLAM where they are separated; voice signals are sent to the switch and data to the Internet (Figure 8.7).

The bit rate on the loop depends on the loop distance. The copper twisted pair cable, because of its impedance, capacitance and inductance, behaves like a low pass filter, and therefore attenuation increases at higher frequencies and at longer distances. Consequently, the longer the loop length (between DSLAM and subscriber), the lower the maximum possible bit rate (Figure 8.8).

Although the DSL technology has been popular since the end of the 1990s, the demand for higher data rate at the loop is continuously increasing. Unfortunately, copper wires cannot deliver over long loop lengths higher bit-rates than DSL rates. The solution to this is fiber optic loops, examined in a subsequent section.

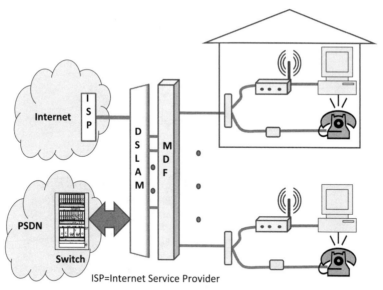

Figure 8.7. Architecture of DSLAM (include MDF).

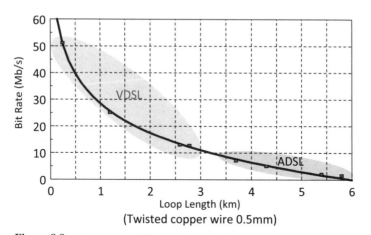

Figure 8.8. Approximate DSLAM bit rates by distance.

8.2.3 Signaling System 7

The Signaling System 7 (SS7) protocol is used to set up and tear down connections across the PSTN network. SS7 was developed to:

- optimize digital network operations,
- be backward compatible with existing switches,

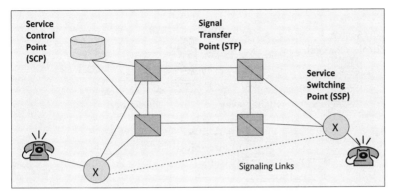

Figure 8.9. Components of the SS7 network.

- meet future requirements, and
- provide a reliable mechanism for data transfer without loss or duplication.

Across the network, nodes that support SS7 are known as signal transfer points (STP); thus, STPs receive SS7 messages from a signaling link and transfer them to another link. STPs monitor messages and maintain connectivity (routing) tables.

There are two STP types, national STPs and gateways. National STPs transfer SS7 messages within the national network. Gateway STPs work with both protocols, national and international, and translate messages from one to the other.

Additionally, service control points (SCP) are computers that maintain databases such as:

- Local number portability (LNP)
- Calling name database (CNAM)
- Home locations register (HLR)
- Line information database (LIDB)
- and more.

Local exchanges are not STPs but service switching points (SSP). SSPs utilize the common channel signaling (CCS)-call processing protocol. Thus, SS7 starts from an end office (STP) and ends at a remote end office (STP) (Figure 8.9).

8.2.4 Getting Connected Across the Network

When the telephone is on-hook, the loop is open and there is no current flow. However, as soon as the telephone goes off-hook, then the loop closes and current flows from a battery connected to it (the loop gauge and length are important as both impact the Ohmic loss and thus the current flow). As soon as this current is sensed by the exchange, it puts a tone on the loop, indicating that the telephone operator can now dial the destination number.

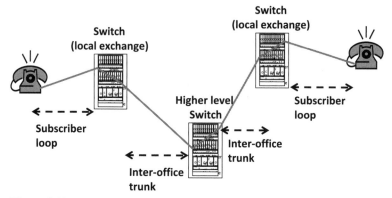

Figure 8.10. Connecting resources across the PSTN network for the duration of a call.

When the destination number is dialed, the originating local exchange (STP) sources an SS7 message to the nearest STP. Each SS7 message contains a label. The portion that is used for routing is known as "routing label;" this includes the destination and the origination points.

The STP consults the data base SCP, and a path from the originating to destination devices is found. Connectivity details (using SS7 messages) are then transferred over signaling links to all nodes on the defined path. The STPs on the path update their connectivity tables and connect a specific input with a specific output of their switch.

If the destination device is not in use, the destination local exchange (STP) puts on the destination loop a ringing tone. When the destination device goes off-hook (it answers the call), the ringing stops and all resources on the identified path are committed for the duration of the call (Figure 8.10).

When one of the two devices goes on-hook, the resources on the path are released and become available for other connectivity.

As a note to this section, local area networks (LAN), although are not operated and managed by local carriers but by private datacom enterprises within a moderate sized geographical area, are also connected with the public network via bridges [19]. As such, the security aspects of LANs also impact the overall security of information and the network.

8.2.5 Network Synchronization

The PSDN network is synchronized to a highly accurate reference frequency of 4 KHz (or 8 KHz). Because the PSDN has a hierarchical structure, synchronization accuracy is relaxed towards the edges of the network.

Traditionally, when the analog signal is converted to digital, a digital clock is required to create and maintain the DS0 rate across the network. In a synchronous network, this clock must meet certain accuracy measures as described by standards.

Table 8.5. Timing accuracy at different layers of the network

Stratum	Minimum Accuracy	Slip Rate
1	10.0^{-11}	2.523 periods/year
2	1.6^{-8}	11.06 periods/day
3	4.6^{-6}	132.49 periods/hour
4	3.2^{-5}	15.36 periods/minute

This clock is generated by a local oscillator which is locked to an accurate reference frequency. This accuracy is required in the multiplexing hierarchy to higher bit rate digital services, DS2 and DS3 (in the U.S.) and also E2 and E3 in Europe, as well as in optical rates that exceed 1 Gbps.

During transmission of electrical and of optical signals, small phase shifts due to parameter and environmental changes cause the payload to be shifted positively or negatively with respect to the receiver system clock; this is resolved by frame resynchronization or by frequency justifications (adjustments) using floating payloads and pointers as in the SONET/SDH [1, 2].

Nodes on the higher level observe the most stringent specification known as stratum 1 (Table 8.5).

In addition to stratum definition with regard to accuracy, network nodes depending on size and function must meet certain accuracy that is specified by standards, which also specify the accuracy of local oscillators or clocks under certain condition, free-running accuracy, hold-over stability and pull-in/hold-in:

- *Free-run accuracy:* the maximum fractional frequency offset that the clock has when there is no reference clock and it is free running at its own accuracy.

- *Holdover stability:* the amount of frequency offset that a clock has after it has lost its synchronization reference.

- *Pull-in/Hold-in:* the clock's ability to achieve or maintain synchronization with a reference. Its range is at least twice its free-run accuracy.

8.2.6 Signal Performance

Performance of the signal is meaningful at the receiver. As the signal propagates over the medium, it undergoes attenuation, pulse deformation, noise addition, and jitter. The end result is a decrease in signal-to-noise ratio (SNR) and the increase in bit error rate (BER). Bit Error Rate (BER) is defined as the number of errored bits to the number of transmitted bits in an interval of time. Additionally, another performance criterion is the error second. The latter is defined as an interval of one second that contains at least one error. And burst errored second (BES) is defined as any errored second that contains at least 100 errored bits (a burst of errors).

These and other performance parameters need be controlled as there are requirements that determine the acceptable limits of performance parameters. For example, BER for telephony (64 Kbps) is expected to be 10^{-3}, for data transmission 10^{-7}, for DS1 and DS3 signals is 2×10^{-10} whereas for optical signals at the Gbps performance is at 10^{-12} and 10^{-14}.

The next chapter takes advantage of the performance parameters of a channel to construct the channel signature used for additional channel authentication, detect signal degradation, and detect channel intrusion, and also to create countermeasure strategies.

8.3 SECURITY ISSUES

As already mentioned, well-known security breaches of wired networks are eavesdropping, illegally placing long distance phone calls "for free," and making phonecalls by placing charges on another subscriber.

Eavesdropping was simple and entailed tapping the loop copper wires. This is because initially the plain old telephone service (POTS) had no means to be authenticated prior to initiating a call. POTS is a passive device that consists of a carbon-based microphone, an inductor-based speaker, a switch to open/close the loop, a hybrid (transformer-based circuit) that converted the two way analog signal flow from/to the loop in a pair of two-wires (one pair to the speaker and the other to the microphone), and a varistor to automatically match the loop impedance (loops are of different lengths) with the hybrid. To date, POTS still exist with the exception that the old rotary dial has been replaced by a two-tone generator key-board (0 to 9, #, *). POTS are not powered by the end user but by the network provider. More complex telephones with memory, display and other features require power by the customer. Today, wired-loop security depends on the security of the neighborhood cabinet from where all loops are distributed to homes, and on the tap-resistance of loop cables from the cabinet to the office; wired cables contain several hundreds of color coded pairs, and the association of colored pair with POTS number is maintained in records that are not public.

Illegal long distance phone calls could be made from a public phone, such as a coin phone. However, this required detail know-how of the signaling protocol and tones (this protocol is not based on packetized messages but on different tones).

Violators who tapped loops or who devised a "smart box" that mimicked signaling tones were identified and apprehended; older vintage movies and shows, such as *Mission Impossible*, have frequently depicted wiretapping and source mimicking. During time, new versions of signaling were defined that were more complex and more difficult to mimic. Although such attacks caused revenue loss to the telephone companies, violators were limited to eavesdropping on conversations and they did not have the means to remotely harvest customer information, alter information, degrade the node's performance, delete the content of nodes and bring the system down as in the Internet. Similarly, SS7 is a robust protocol that flows in small messages over dedicated links. Tapping SS7 links to read SS7 messages is complex and not rewarding to the violator, and there has not been a vecent public announcement (to this author's knowledge) that such a violation has taken place.

DSL loops that connect SOHO are not immune of security issues. Because the digital data path is connected to the Internet network, then the security issues that already have been described hold, including VPN security issues; that is, virus, worms, Trojan horse, etc.

To protect the computer, a firewall may be placed between the DSL modem and the computer, in the order DSL–Firewall–PC.

Additionally, when the WiFi interface between the DSL modem and the computer is used, the same WiFi security risks hold as also described.

8.4 SECURITY COMPARISON BETWEEN PSDN AND IP

The synchronous PSDN network can be contrasted with the asynchronous data network, and particularly with the Internet.

PSDN is based on standardized synchronous multiplexed frames, such as DS1/E1 and SONET/SDH rates [1], which are circuit switched; the connection through a switch has been established during the call initiation procedure or by node provisioning.

A fundamental difference between PSDN and IP is that the latter transports packet-size information at asynchronous rates, whereas the former transports byte-size information in a continuous, repetitive and synchronized manner with strict throughput and travel delay constraints.

When packets enter a data node or router, they are stored in memory until they are routed to an output buffer. That is, the PSDN provides a hard-wired connection between input–output whereas the router does not, and so the data network is also called "connectionless;" the route of connectionless networks are not decided by a centralized network controller but is determined according to a router algorithm. It is the storing of data in routers that allow smart but malicious programs to creep into the computer; when these programs are executed, one of many undesirable actions takes place such as spoofing, cloning, file deletion, file copying, data harvesting, and so on.

Additionally, provisioning and software upgrades of PSDN nodes are accomplished without service interruption and over an overlay network using a proprietary Intranet, proprietary and secure tunneling, or memory hard-disks that only authorized and trained technical personnel handles. In contrast, IP networks need to interrupt service in order to upgrade a router with new software.

Finally, harvesting one's information in the traditional wired PSDN network requires intervention of the physical medium, whereas in the IP network information may be harvested from a remote computer.

REFERENCES

1. S.V. KARTALOPOULOS, *Understanding SONET/SDH and ATM*, IEEE Press, 1999.
2. S.V. KARTALOPOULOS, *Next Generation SONET/SDH: Voice and Data*, IEEE/Wiley, 2004.
3. S.V. KARTALOPOULOS, *Introduction to DWDM Technology: Data in a Rainbow*, (Published in US and in India) Wiley/IEEE Press, 2000.
4. S.V. KARTALOPOULOS, *Fault Detectability of DWDM Systems*, IEEE Press, 2001.
5. S.V. KARTALOPOULOS, *Introduction to DWDM Technology: Data in a Rainbow*, (Published in US and in India) Wiley/IEEE Press, 2000.
6. B. BOSIK and S.V. KARTALOPOULOS, "A Time Compression Multiplexing System for a Circuit Switched Digital Capability", IEEE Transactions on Communications, vol. com-30, no. 9, September 1982, pp. 2046–2052.
7. S.V. KARTALOPOULOS, "The VLSI Chip of the Time Compression Multiplexer for the Circuit Switched Digital Capability", Globecom 82, Miami, FLA, November 1982 pp. 386–389
8. S.V. KARTALOPOULOS, "Circuit Pack for Digital Loop Carrier Transmission System", US patent issued no. 5,383,180, January 17, 1995.
9. ITU-T Recommendation I.432, "B-ISDN User-Network Interface-Physical Network Specification".
10. ITU-T Recommendation I.580, "General Arrangement for Interworking between B-ISDN and 64 Kb/s based ISDN", Dec. 1994.
11. ITU-T Recommendation Q.931, "ISDN UNI Layer 3 Specification for Basic Call Control", 1993.
12. ITU-T Recommendation G.991.1, "High bit rate Digital Subscriber Line (HDSL) transceivers", Oct. 1998.
13. ITU-T Recommendation G.991.2, "Single-Pair High-Speed Digital Subscriber Line (SHDSL) transceivers", Feb. 2001 and Amendment 1 (11/2001).
14. ITU-T Recommendation G.992.1, "Asymmetrical Digital Subscriber Line (ADSL) transceivers", Jul. 1999, and Annex H (10/2000) and Corrigentum 1 (11/2001).
15. ITU-T Recommendation G.992.2, "Splitterless Asymmetrical Digital Subscriber Line (ADSL) transceivers", Jul. 1999.
16. ITU-T Recommendation G.993.1, "Very high-speed Digital Subscriber Line foundation", Nov. 2001.
17. ITU-T Recommendation G.995.1, "Overview of Digital Subscriber Line (DSL) Recommendations", Feb. 2001, and Amendment 1 (11/2001).
18. ITU-T Recommendation G.1010, "End-user multimedia QoS categories" Nov. 2001.
19. FIPS PUB 191, Guideline for the Analysis Local Area Network Security, November 9, 1994.

Chapter 9

Network Security: Optical Systems

9.1 INTRODUCTION

The copper-wire backbone network was unable to meet the bandwidth demand as a result of new services and thus it quickly became unsuitable and uneconomical. The medium that offered practically unlimited bandwidth, as demonstrated in the 1980s and particularly in the 1990s, was the fiber optic silica-based medium. As such, copper-based trunks were quickly replaced by fiber, the transmission characteristics of which were also improving with time.

In addition to the demand for more bandwidth over the network, there were several other factors that drove the copper replacement by silica fiber:

- Silica in raw form is found in abundance on earth. It is found in sand (SiO_2), and also in organic matter (rice, corn). Therefore, it is an inexpensive material.

- The propagation characteristics of silica-based fiber are stable over many years. Certain elements that are used to make the communications fiber with desirable propagation characteristics (fluoride, boron, erbium, aluminum, etc.) are also found relatively easily.

- Very accurate methods have been developed to manufacture fiber meeting the ITU-T strictest requirements and tolerances. ITU-T specifies a core diameter of 50 μm for multimode fiber and of 9 μm for single mode fiber [1]. Similarly, the outside diameter of both multimode and single mode is specified at 125 μm. These accuracies must be maintained over the length of unspliced fiber, which is many kilometers.

- Specialty fiber has also been manufactured to meet specific requirements; dispersion shifted, dispersion compensated, water-free fiber, and many more [2–7].

- A large variety of components (optical and photonic, fixed and tunable) have been developed to construct an all-optical node and an all-optical link. Optical

Security of Information and Communication Networks, by Stamatios V. Kartalopoulos
Copyright © 2009 Institute of Electrical and Electronics Engineers

switches, filters, multiplexers/demultiplexres and optical amplifiers are among them [8].

- Fast switching modulation methods have been devised that can generate bit rates beyond the 40 Gbps mark.

- A large variety of protocols for traffic control (route discovery, traffic balancing, congestion avoidance, etc) have been developed.

- A new technology known as wavelength division multiplexing (WDM) has been developed. This multiplexes many optical channels in a single fiber, the grid of which has been standardized for dense channels (DWDM) and for coarse channels (CWDM) [9]. Imagine, 1000 channels at 10 Gbps produce an aggregate traffic of 10 Tbps per single fiber, and there may be 1000 fibers in a single cable. This constitutes a bandwidth as an almost inexhaustible medium, unmatched by any other practical medium known today.

- Finally, the glass-based fiber medium is immune to radio interference, has the tensile strength of steel or better, and (with equipment and personnel training) is relatively easily spliced and connectorized.

As a result, the backbone network in several continents is mostly fiber-based, with WDM the technology of choice for both synchronous payloads (voice, real-time video) and asynchronous payloads (high speed data, Internet). However, fiber transports a huge bandwidth, a good part of which contains extremely sensitive information such as financial and banking, medical, personal data, credit card numbers, and classified documents; it is this type of information on the network that wets the appetite of intruders and of others who enjoy cyber-war tactics and like to cause network congestion, denial of service, derailing information, and more. Therefore, security issues are not absent from optical networks, which is the object of this chapter.

This chapter assumes that sensitive information is already encrypted by one of the well-known methods already described. In addition, it assumes that an intruder attempts to "tap" the physical medium, the fiber. Therefore, this chapter examines the vulnerabilities of optical networks, describes a method by which medium intrusion is detected, and describes counter-measure strategies, network self-defense, and network counter-attacking scenarios. Additionally, this chapter describes the security aspects of free space optical (FSO) networks; distinguishes FSOs as fiberless terrestrial optical networks (FtSO) and free space satellite optical networks (FsSO); although both use laser beams and no fiber, the transmission characteristics, ambient environment and security issues are very different.

9.2 LAYERS OF NETWORK SECURITY

Optical synchronous network nodes do not store customer information as data networks do. Therefore, an intruder who would like to gain access to a synchronous node has two opportunities: tap the fiber and eavesdrop or mimic network management, which is over a secure Intranet with better security than the Internet. A third opportunity is to gain physical access to a node in a central office or a management center, both of which are secure buildings and only personnel with proper authorization can enter. As a result, three major areas in communication networks require security protection: the physical layer with intrusion detection and countermeasures, the MAC/Network layer, and the Information (customer data) layer.

- If the physical layer of the communications network (at the link or node) is tapped to eavesdrop, using cryptanalysis and superfast computation, the cipher key may be found to decipher encrypted messages, to alter message and re-inject messages. Tapping the fiber can be accomplished in a neighborhood cable distribution enclosure, or in an unprotected area through which the fiber runs. Information extracted from it is difficult but is not impossible. Because the medium runs for many kilometers across unprotected areas, *network physical layer security* is accomplished by intelligent detecting mechanisms.

- Gaining control of a node via the management port is an extremely difficult task, as only certain skilled and authorized personnel can do so. Gaining control of the management channel requires specialized knowledge and therefore it may be difficult but is not impossible. Therefore, *security on the MAC/Network layer* is accomplished by intelligent detecting mechanisms and by strong encryption of messages.

- The security and privacy of customer data is accomplished at computer-based end-devices, regardless if this is synchronous or asynchronous. To accomplished this, strong scramblers/descramblers or strong encryption algorithms are used.

In summary, an intelligent network should not only protect data by using encryption at the information layer (end-to-end) but also protect itself at the control and at the physical layers. Securing data at the physical layer is an engineering task that employs mathematical analysis to design sensitive and reliable detectors. Securing data at the Information layer is a task closer to mathematics. Securing data at the MAC/Network layer employs both mathematics for encryption and engineering for intrusion detection.

9.3 SECURITY OF OPTICAL ACCESS NETWORK

9.3.1 Passive Optical Networks and Fiber-to-the-Premises

Optical access networks such as passive optical networks (PON), also known as fiber to the premises (FTTP), fiber to the home (FTTH), fiber to the enterprise (FFTE), fiber to the curb (FTTC), fiber to the telecom enclosure (FTTE), fiber to the cable distribution (FTTD), and so on, have an open structure, serve many end users, transport sensitive end-customer information, and are candidates to become the target of bad actors who will be looking for vulnerabilities. In the literature, the acronym EPON is a PON that has adapted the Ethernet protocol, GPON is a PON at Gigabit rates, and so on. Herein, for simplicity and to avoid confusion, all PON variants are named as PON and all fiber-to-the-x as FTTx, and these two terms, PON and FTTx, are used interchangeably.

The major components of PONs (from the central office to the end-user) are an optical line termination (OLT), an optical network unit ONU), and many network terminals (NT) (Figure 9.1).

The PON structure is open because the ONU is located in an enclosure outside the central office (CO) and most likely in the vicinity of a neighborhood, whereas the NT is often placed right at the customer premises or near it. The OLT is most likely housed inside the CO. Although these two components, ONU and NT, are under the control of the network operator, it is possible that an intruder may penetrate the ONU or the NT enclosure. In addition, sophisticated attackers may gain access to the fiber medium and tap it (the fiber can be several kilometers long). Therefore, a PON that serves many hundreds of end-users, some of which may be non-cooperative, may be a target.

Typically, there are two fundamental PON or FTTH architectures. One PON adapts WDM technology (WDM-PON) and thus has many optical channels; another uses a single optical channel and another adapts time division multiplexing technology (TDM-PON) and has a single optical channel. Each of the two has weaknesses and strengths in a complementary manner [10].

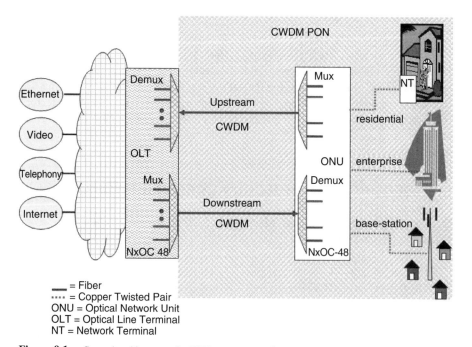

Figure 9.1.   General architecture of a PON access network.

- The WDM-PON has many optical channels multiplexed at the OLT, which are transmitted to the ONU where there is an optical multiplexer. From the ONU, each optical channel is sent to the NT (Figure 9.2). Thus, although the deliverable bit rate is high, up to 10 Gbps, the number of end-users is small, as many as the optical channels. However, the WDM, depending on fiber length between OLT and ONU, may not require optical amplification, and uses mostly well-established optical (mostly passive) components.

- The TDM-PON uses a single optical channel, the power of which is split in several prongs (Figure 9.3). This implies that at the splitter (before or after it) an amplification strategy must be adapted, such as a semiconductor optical amplifier (SOA) or a Raman amplifier after the splitter. After the splitter, the ONU has a complex design because it requires synchronized time slot recognition; time slot assignment is accomplished via ONU provisioning. Each NT is allocated a specific time slot, but the fiber length between the NT and ONU is unequal; thus the propagation time. Therefore, time slot synchronization (by time slot buffering) is required, which is particularly complex because optical signal buffering is cumbersome if one considers that temperature variations affect the propagation time and cause uneven delays. Moreover, each NT must have a complex design so that it can extract from the series of time slots the one it belongs to. In the upstream direction, this is also complex because the NT must insert information exactly in its own time slot; if inserted a little sooner or later, data will conflict with one of the adjacent time slots.

A third architecture has been developed that takes into account the pros and cons of the two aforementioned ones. This is called hierarchical WDM/TDM (hCD-PON) [11–13] (Figure 9.4).

The hCT-PON uses the WDM technology between the OLT and ONU, and the TDM between the ONU and NT. At the ONU, each optical channel is organized in

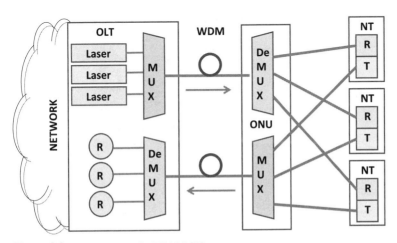

Figure 9.2. Architecture of a WDM-PON.

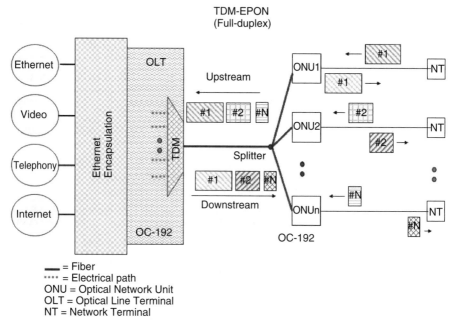

= Fiber
= Electrical path
ONU = Optical Network Unit
OLT = Optical Line Terminal
NT = Network Terminal

Figure 9.3. Architecture of a TDM-PON.

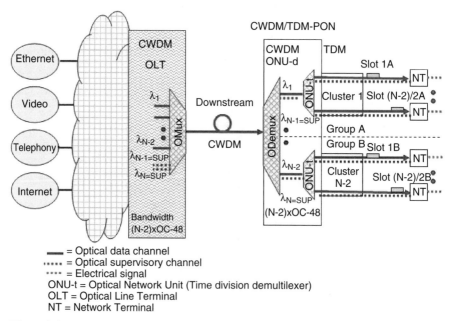

= Optical data channel
= Optical supervisory channel
= Electrical signal
ONU-t = Optical Network Unit (Time division demultilexer)
OLT = Optical Line Terminal
NT = Network Terminal

Figure 9.4. Architecture of the hCT-PON in the downstream direction.

clusters that serve NTs of small distance variability and thus have small delay variability, although the average variability from cluster to cluster may be significant. Thus, using the same fiber length (the one that reaches the furthest NT) within a cluster of NTs, the delay variability is eliminated in both directions, downstream and upstream. Additionally, a guard-band at either side of the time slots accounts for unforeseen time variability due to jitter and other causes.

Assuming that each optical channel delivers 2.5 Gbps to each NT, and that each end-user on the average requires 2.5 Mbps, each cluster can be fanned out to 1000 NTs. The rate of 2.5 Mbps is the most popular and low-cost bit rate to date; 10 Gbps or 10 GbE is equally possible as there is no practical limitation in the method. In the upstream direction, a similar process takes place in reverse; however, time slots are self-synchronized by virtue of equal-distant fibers in the same cluster (Figure 9.5).

The hCT-PON in its initial definition was engineered using the coarse ITU-T G.694.2 WDM grid, which consists of eighteen optical channels (Table 9.1). Two channels of them (with 1 + 1 protection) have been allocated for supervision, management, control and security, and sixteen channels for customer data. The sixteen optical data channels are divided in two groups of eight. Each optical channel carries 1000 data time slots (Figure 9.6), each time slot carrying 2.5 Mbps; thus, the hCT-PON is able to serve from a single OLT 16,000 NTs or end-users at 2.5 Mbps each. Notice that 1000 time slots on a cluster repeat every 125 msecs.

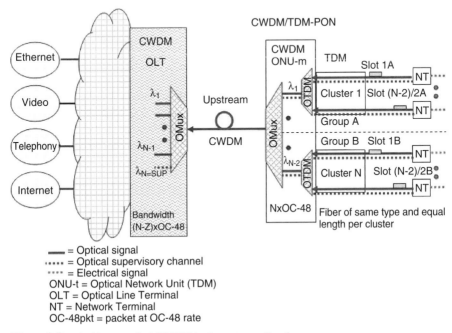

= Optical signal
= Optical supervisory channel
= Electrical signal
ONU-t = Optical Network Unit (TDM)
OLT = Optical Line Terminal
NT = Network Terminal
OC-48pkt = packet at OC-48 rate

Figure 9.5.   Architecture of a hCT-PON in the upstream direction.

Table 9.1. CWDM center wavelengths of the ITU-T grid

Wavelength (nm)	Frequency (THz)	Wavelength (nm)	Frequency (THz)
1271	232.6	1451	206.5
1291	230.1	1471	203.6
1311	227.7	1491	200.9
1331	225.1	1511	198.2
1351	221.8	1531	195.6
1371	218.5	1551	193.1
1391	215.4	1571	190.7
1411	217.3	1591	188.2
1431	209.3	1611	185.9

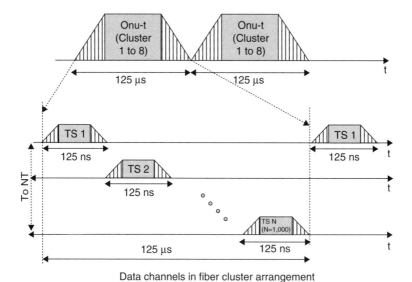

Data channels in fiber cluster arrangement

Figure 9.6. Time slotted data channels in the hCT-PON.

The two hCT-PON supervisory channels are also time-slotted with guard-bands so that there is a time slot per NT; this allows for each NT to be individually supervised and controlled by the OLT.

Because the bandwidth granularity of hCT-PON per NT is 2.5 Mbps, if an NT requests more bandwidth then more than one time slot may be allocated dynamically to each NT. Thus, an NT may be at 2.5 and another at n × 2.5 Gbps. Additionally, the signal over a time slot may also be further multiplexed to include voice (64 Kbps) and Internet data commensurate to DSL data rates; notice that the traditional clock of 8 KHz is still maintained for compatibility with the existing optical synchronous networks. Thus, the hCT-PON is backward compatible with existing technologies, yet it is flexible to offer an elastic bandwidth in 2.5 Mbps increments. Moreover, if the 2.5 Mbps carry DS1 (1.544 Mbps) or E1 (2.048 Mbps) signals, then the 2.5 Mbps may be further demultiplexed to support POTS or ISDN services only, if needed.

An additional feature of the hCT-PON is that it can support simultaneously two topologies, which are fundamentally different: from the ONU to NTs a cluster may support a star topology as described, or it may support a linear topology for LAN/MAN applications (Figure 9.7).

The hCT-PON is not limited to bandwidth or customer connectivity. When a coarse 36 channels grid is used (although not ITU-T specified, yet), more than twice as many NTs can be served, or more bandwidth can be delivered, or a combination of both. If one of the ITU-T G.652 dense WDM grids is used, again more bandwidth and more NTs may be served. In any case, the choice depends on the specific application, on engineering and the economic model [14–16].

In terms of security, because of the time slotted supervisory structure for each NT, the OLT authenticates all downstream components, ONU and NTs; NTs are authenticated when they request service, one at a time or simultaneously. In addition, a secure and dynamically changing protocol may be used to deliver time-dependant keys, time-dependant time slots, etc. [17, 18]. Such flexibility in bandwidth deliverability, number of end-users and network security has not yet been encountered in other PON derivatives. In addition, other improvements are under investigation in order to enhance security with existing protocols [19].

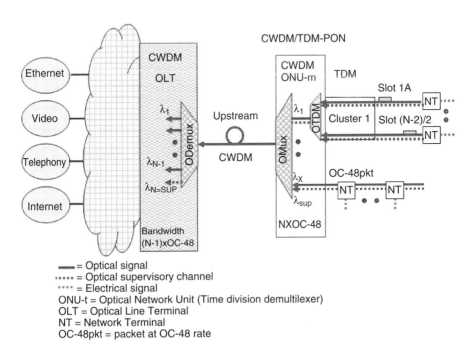

Figure 9.7.   The hCT-PON supports simultaneously star and linear topologies.

9.3.2 Security and Vulnerabilities of PONs

In all optical networks, the fiber medium is easier to attack as it is impossible to be guarded for many kilometers. Thus, the fiber link between an OLT and ONU is vulnerable to tapping. The problem with all access networks including PONs is that they are an almost oxymoron: they should not be complex, they should have a low capital expense and low-cost operating expense, and yet be very robust, reliable, flexible, and secure.

Regarding security of optical access networks (PON), the link between OLT and ONU is vulnerable to tapping. A WDM-PON carries many optical channels in this link and thus specialized instrumentation to tap each channel is required. In contrast, a TDM-PON carries a single optical channel and is more vulnerable. The hCT-PON also carries many optical channels and two supervisory, but each channel is time division multiplexed. Thus, the hCT-PON is much more difficult, because in addition to optical channel selection, time slot synchronization is required. This, along with dynamic provisioning and dynamic time slot allocation, makes the hCT-PON OLT-ONU link more secure [20, 21].

The segment between the ONU and NT is more vulnerable in the WDM-PON because it carries a single channel. Tapping the TDM-PON fiber requires specialized equipment that can be synchronized with the synchronization scheme TDM-PONs specify. The hCT-PON is much more difficult, because in addition to time slot synchronization, the supervisory channel time slot needs to be synchronized. This, along with dynamic provisioning and dynamic time slot allocation, makes the hCT-PON more secure.

An additional security feature of the hCT-PON is the authentication of ONU and NT, reducing the risk of source mimicking or masquerading. Moreover, if the channel signature method (described in a subsequent section) is used, then authentication of ONU and NT is enhanced and physical attack may be identified. In the latter case, the OLT may autonomously re-allocate timeslots to a suspicious ONU-NT link.

9.3.3 Types of Attacks In Optical Access Networks

9.3.3.1 Interception of the Physical Plant

Interception refers to attacking the physical plant, the fiber. In all PONs, the link OLT-ONU, as well as the links ONU-NTs are vulnerable to interception. The cabinets or enclosures in the outside plant are supposed to be protected by entry codes and by intrusion detectors.

In fact, enclosures that contain cryptographic modules, must meet four security levels [22–26]:

- *Security Level 1 (SL-1)* specifies basic requirements to software and firmware components that are executed on an unevaluated general purpose computing environment. No specific physical requirements are provided for a SL-1 module. When performing physical maintenance, all plain text secret and private keys and other *critical security parameters* (CSP) in the cryptographic module shall be zeroed; this is performed procedurally by an operator or autonomously by the cryptographic module.

- *Security Level 2 (SL-2)* protects against unauthorized physical access by adding the requirement for tamper-evidence coatings, seals, and locks on door and covers. At this level, when an operator needs to gain access to perform a set of services, the operator requests authorized access and the cryptographic module authenticates the authorization access. Thus, SL-2 requires at minimum role-based authentication.

- *Security Level 3 (SL-3)* attempts to prevent an intruder from gaining access to CSPs in the cryptographic module. SL-3 requires physical security mechanisms that detect attempts to physical access as well as reporting mechanisms. When the enclosure is violated it should cause serious damage to the cryptographic module. Thus, SL-3 requires identity-based authentication mechanisms. Circuitry detects an attempt to open the cover or the door of a module or to access the maintenance access interface. Then, the circuitry zeroes the plaintext secret and public cryptographic keys and CSPs contained in the cryptographic module.

- *Security Level 4 (SL-4)* provides a complete envelope of physical security around the cryptographic module and it detects any attempt of penetrating the cryptographic module and gain access of the cryptographic keys. SL-4 also protects against security compromise as a result of external environmental avert conditions and fluctuations. Thus, the cryptographic module contains either *environmental failure protection* (EFP) or *environmental failure testing* (EFT) mechanisms. Attempting to remove or dissolve protective coatings should result in serious damage of the cryptographic module.

Therefore, in PONs that are used in very secure environments either the complete set of aforementioned requirements is needed, or a subset that pertain to enclosure penetration (SL-2 and SL-3 or both). Because ONUs most likely do not contain cryptographic keys, it may not be necessary to comply with the full SL-4 requirement, although this needs to be evaluated for NTs if they do.

9.3.3.2 Eavesdropping

Eavesdropping in PON networks is the simplest type of physical attack. In PONs, the link between OLT-ONU is from few to several kilometers long, whereas the ONU and the NT are in the outside plant. Thus, the fiber is more vulnerable for tapping and eavesdropping because the cabinets are supposed to include sensors that detect intrusion and unauthorized entry. Nevertheless, the eavesdropper must have thorough knowledge of optical channels (in WDM-PON and hCT-PON), data time slot assignment (in TDM-PON and hCT-PON), and supervisory time slot assignment for each end-user (in hCT-PON) in order to access a user channel [27].

9.3.3.3 *Source Mimicking and Theft of Service*

NTs in the WDM-PON are serviced by a single optical channel. Thus, it may be viewed as a point-to-point single wavelength connection. Security of the link is as good as the authentication protocol.

NTs in the TDM-PON are serviced by specific time slots. If the time slots are fixed, then it is relatively easy to identify the association time slot–end-user. When this is accomplished by a bad actor (Evan), security of the slotted channel is as good as the authentication protocol.

NTs in the hCT-PON are grouped by clusters. NTs within the same cluster share the same wavelength in both upstream and downstream direction and each NT has its specified time slot, in which the NT sends and receives data. Nevertheless, the time slot duration and sequence is under OLT control and it may change dynamically upon user bandwidth request. Now, if Evan wants to impersonate or mimic a user (or NT), he must have knowledge of the bandwidth and service level agreement (SLA) requested by the user under attack and granted by the OLT, and also knowledge of the NT identification. Moreover, Evan should be able to defeat the authentication protocol and the secret key defined by OLT during the key distribution process when service is granted to a NT. Therefore, source mimicking and theft of service is much more difficult compared with both WDM-PON and TDM-PON.

9.3.3.4 *Attacking the Supervisory Channel*

In hCT-PON, the supervisory channel transports operation, administration and maintenance (OAM) control messages. It may also carry scrambling parameters and secret keys to NTs, user identifier, and username/password, or it may contain important NT configuration data, and so on.

Thus, attacking the supervisory channel may be more serious than attacking the data channel in terms of denial of service. In view of this, the NT and OLT periodically exchange encrypted control messages that are embedded in data packets confirming their proper operation. In addition, more sophisticated countermeasure mechanisms may be employed that can readily detect intruders by monitoring the channel signature.

9.3.3.5 *Security Measures*

Reports of malicious attacks such as eavesdropping, data theft, identity theft, bank account theft, and so on, in wired, wireless and data networks are not uncommon. Although the PON requires sophistication and expertise, bad actors are also sophisticated and with the proper know-how they will manage to get a handle on it. Therefore, PONs should have built in security and countermeasure features from the outset, such as:

- End-user terminal authentication. Upon activating the NT, an NT code is sent to the network which authenticates it and grants service. In view of possible attack, the data time slot may be reassigned; that is, the hCT-PON OLT in particular may dynamically reassign the time slot to NT upon activation and authentication, so that it is very difficult for Evan to track the time slot length and time slot association with the end-user.

- NT authentication. As the network expands and new NTs are added to the network, self-discovery takes place to validate the NT type and assign an ID to it. For security purposes, the NT ID code may randomly change, thus diminishing the probability of NT impersonation or mimicking. This is easily done in hCT-PON since the NTs are under OLT control.

9.3.4 Detecting Fiber Intrusion and Countermeasure Strategies

This is addressed with well-known methods that may be incorporated in the OLT so that it can detect fiber intrusion. This method will be further elaborated upon in a subsequent section in this chapter. However, suffice it to say that when intrusion is detected, then the OLT activates a countermeasure strategy whereby it sends client data over a different time slot, whereas it continues transmitting decoy messages over the attacked time slot.

Another deterrent strategy is key renewal; the keys are frequently updated so that the attacker cannot obtain sufficient data to decrypt the cipher text. Key distribution protocols may also be employed in hCT-PON, and with frequently renewed keys the possibility of a successful security breach becomes negligible [28].

9.3.5 Backbone Optical Networks

Backbone optical networks consist of switching nodes or routers of high data capacity; nodes are within central offices that are not accessible by non-authorized personnel, and thus gaining unauthorized access to a node is not an easy matter. In fact, unauthorized access becomes even harder if nodes become all optical, in which case encrypting the data stream is in the optical regime [29]. However, nodes are interconnected with DWDM fiber links. Each link is many kilometers long and carries a huge amount of data that may be in excess of Tbps [30].

It is the huge amount of data in a single fiber that whets the appetite of the sophisticated intruder. Tapping the fiber is not trivial but it is not impossible. A sophisticated Evan can use the self-slicing tools that cut the sheath of the fiber cable, isolate the fiber in question and tap it without disrupting service; thus, tapping the cable goes unnoticed. Then, he proceeds to copy data using portable equipment, which under the scrutiny of a supercomputer attempts to crack codes and decrypt cipher texts [31, 32].

Because of the long fiber length and the inability to guard it, the only defense in this case is detecting the tapping of the fiber. This is accomplished by continuously monitoring the performance signature of each channel in the fiber [33]. When tapping the fiber, a small amount of power is extracted, which with the right method at the receiver can be detected as soon as tapping occurs. Using an additional sophisticated method, the point of tapping may also be calculated. In this case, personnel may be dispatched to the point of intrusion and/or a countermeasure strategy may be employed. Moreover, it is possible that the network itself may employ a counter-attacking strategy, as will be described in this chapter.

9.4 CYBER-ATTACK DETECTION MECHANISMS

The physical network could be attacked by sophisticated intruders at the node, at the outside plant cabinet, or at the fiber transmission medium.

Network nodes are in well-secured central offices, and therefore this section does not examine node intrusions. If it occurs, however, it requires some internal cooperation, and this topic is outside this text's scope. At the outside plant, if cabinets handle keys and cryptographic algorithms, then they should comply with the FIPS PUB 140–2 and Annex A-D or part thereof. The third element that presents an opportunity to Evan, the cyber-attacker, is tapping the transmission medium, the optical fiber.

The fiber is a medium that transports modulated laser light and has been through a number of optical and photonic components over the laser-to-photodetector link. All components on the link, including the fiber itself, have specific characteristics that are stable over relatively long periods, and if they change, they do so gradually; gradual changes may occur either because of environmental changes (temperature) or because of natural degradation of components. As a consequence, the transmission properties of modulated light in fiber, including optical power, noise, jitter, pulse shape, and so on, are affected by the link in a unique way that establishes a *performance vector* or a *channel signature*. It is this unique channel signature that can be exploited to authenticate the link (including the source) and also detect an attack (tapping) on the medium as it occurs, since the attack changes some of the parameters of the propagating light in the link.

Subsequent sections outline this methodology, which detects at the receiver such an attack in real time and in service. A thorough analysis of this methodology is beyond the scope of this book as its objective is link security and not performance of optical communications (analysis of the method can be found in [34]). However, subsequent sections also architect an integrated low-cost circuit that can be embedded in the receiver; this along with a protocol is also used for channel protection, channel equalization, and particularly for security countermeasures and also for counterattacks.

9.4.1 Channel Performance Vector as Channel Signature

In optical communications, as light propagates in the optical medium, a light-matter interaction takes place that affects the propagation characteristics and the photon characteristics (such as state of polarization (SoP)). These interactions affect the signal performance when the signal arrives.

The factors that influence the signal performance are linear and non-linear. Among them are:

- Attenuation
- Cross-talk
- Dispersion and dispersion slope
- Dispersion Group Delay (DGD)
- Polarization mode dispersion (PMD)
- Polarization dispersion loss (PDL)
- Four wave mixing (FWM)
- Cross-phase modulation (XPM)
- Self phase modulation (SPM)
- Modulation instability (MI)
- Inter-symbol interference (ISI)
- Noise and jitter.

These are analytically known for a particular link and channel, and the signal performance parameters form a vector that constitutes the channel signature ID [35–45]. How these parameters affect the signal performance is identified in Table 9.2 [46].

Based on this table, key observable signal performance parameters for a fiber link are:

- noise floor,
- ringing,
- jitter,
- bit error rate (BER),
- signal to noise ratio (SNR),
- min-max signal power levels, and
- pulse shape distortions (skew, kurtosis).

From them, the Q-factor, the noise factor (NF) is calculated. These parameters constitute the optical channel performance vector $P\{x\}$, and as parameters change, so does the performance vector.

In this method, during link set-up the initial performance vector $P(x, t_0)$ constitutes the initial signal signature or channel ID and its performance parameters that constitute the performance vector are stored in a register. Thereafter, any parameter change for whatever reason will affect $P(x, t_0)$, or the channel ID. However, in order to track vector changes, it is necessary that the performance vector $P(x, t)$ is frequently monitored without service disruption and is compared with the initial

Table 9.2. Effect on signal performance when parameters vary

PARAMETER	EFFECT ON SIGNAL
Optical signal power level:	Reduces eye opening; increases SNR and BER
OCh center frequency deviation:	Reduces eye opening; increases BER; increases cross-talk
OCh width (broadening):	Reduces eye opening; increases SNR and BER; increases cross-talk
OChs (spacing) separation:	Reduces eye opening; increases BER (combined effects of items 2 and 3)
Dispersion (chromatic, polarization):	Eye-closure due to chromatic dispersion; eye closure due to PMD induced DGD; increases ISI, SNR and BER
State of polarization instability	Increases BER; increases jitter/wander
Modulation depth (peak to valley):	Reduces eye opening; increases BER
Modulation stability (peak and valley):	Reduces eye opening; increases BER
Signal echo and singing:	Adds to laser chirp, to signal jitter and noise
Fiber dielectric non-linearity:	Eye-closure due to DGD caused by SPM, induces XPM, SPM and FWM on multiplexed opt signals.
Fiber length	Attenuation increases by length
Fiber type	Affects dispersion, polarization, attenuation constant, propagation characteristics, frequency band selection

$P(x, t_0)$. Doing so, the method cannot tell exactly which parameter changed but it provides a bulk indication of the type of change and particularly of the rate of change (that is, how fast the change occurred). This provides the forensics for root cause analysis and helps to discriminate between degradation, failure or malicious attack.

9.4.2 Discriminating Between Faults and Attacks

In general, to discriminate between faults or degradations and intrusion requires a network that monitors the signal performance vector in real-time and in-service at each receiver [47–49]. Current in-service methods are limited to estimating BER as part of the error detecting and correcting (EDC) mechanism embedded in data. However, the detecting/correcting capability of EDCs is limited (strong EDS detect up to 16 errored bits and correct 8), and this method is relatively slow (reliable BER estimation requires many frames). For example, at 10 Gbps and for an expected performance 10^{-15} BER, may be required to 26 hours for reliable BER estimation, or 1.7 min for 10^{-12} BER.

However, the channel performance is not only affected by intrusions but also by degradations and by component failures [50, 51], particularly by inadvertent fiber cuts. Thus, in order to distinguish between a degradation/failure and intrusion, the following pragmatic assumptions are made (and also the action that needs to be taken) [52–56], (Figure 9.8):

- Component degradation takes place slowly, and so is the rate of performance change of one or more optical channels. This is detected by periodically monitoring the performance vector, and by performing a time-base analysis; that is, compare current with previous performance vectors to determine the rate of change.

- Component failure is abrupt and affects a single optical channel or a group of channels, causing permanent disruption of service to affected channels. This is detected by performing a time-base & analysis and also by correlating reports from receivers; the failure may be localized by other reports from distributed detectors strategically placed on the link.

- Intrusion causes an abrupt performance change in one or more channels but does not cause disruption of service. The degradation, depending on tapping method, may be as small as 0.5 dB (this corresponds to the value of a simple fiber tap loss) and as fast as a few microseconds. Intrusion is detected by performing a time-base analysis and also by correlating reports from receivers; the intrusion point may be localized by other reports from distributed detectors strategically placed on the link.

In Figure 9.8, two performance threshold levels are used: the severe degradation is unacceptable and is considered by the network as a permanent failure (in which case the unit with the fault is replaced). The second degradation threshold level (above the severe degradation) is optional and is only defined to proactively perform switch-to-channel protection, when the rate of degradation is faster than anticipated; it compensates the switch to protection protocol execution time prior to crossing the severe threshold. This way, service is provided, meeting the performance criteria.

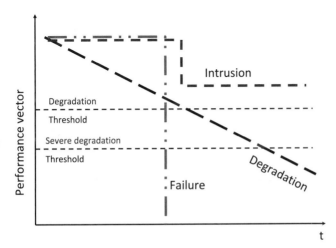

Figure 9.8.   Profile of link failure, degradation and fiber intrusion attack.

For example, ITU-T Recommendation G.821 (1996) for digital transmission systems and for rates below the primary recommends performance thresholds, based solely on BER, as follows:

- BER performance is *acceptable* if it is better than 10^{-6} for more than 1 sec,
- BER performance is *degraded* if it is between 10^{-6} and 10^{-3} for any 1 sec period,
- BER performance is *severely errored* beyond that point and for 1 sec, and
- BER performance is *unavailable* if the severely errored performance persists for more than 10 consecutive seconds.

9.4.3   Method for Estimating the Performance Vector and Channel ID

Since the invention of the cathode ray tube, a traditional and well-established method in channel performance evaluation of communications links has been the eye diagram. The eye diagram is the result of superpositing periodic pulses on the screen of an oscilloscope (Figure 9.9). The amount of eye closure or openness is a bulk indication of the signal performance [57].

To date, this method uses sophisticated high speed sampling oscilloscopes to capture eye diagrams at line rates that exceed 1 Gbps. Such instruments are costly, relatively bulky for use at each input port of a communications node, and disrupt service in order to perform the evaluation tests. Nevertheless, channel and link evaluation tests using such instruments are important and accurate, and are used prior to establishing service over the channel/link; that is when the link is out of service. The observable quantities using sampling oscilloscopes is the eye-diagram closure based on specific binary patterns that have been injected at the source. Thus,

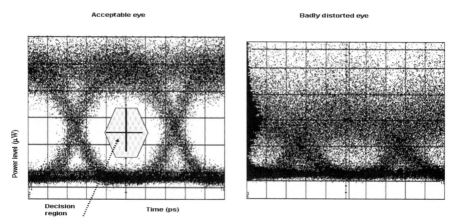

Figure 9.9.   Sampled eye diagrams for the evaluation of channel performance. The more open the eye the better the signal performance.

comparing the transmitted with the received pattern, an accurate calculation is made on the bit error rate, whereas SNR calculations are made directly from the eye-diagram at the receiver. When this test is completed, the instrument is disconnected from the receiver.

This section attempts to answer the obvious question:

Is there a method by which the channel/link performance parameters can be estimated in real-time and in-service (without service disruption)?

This author believes that the answer is "yes". A thorough analysis of the estimation method will not be repeated here, as it can be found in the literature [34, 58]. However, this section lists the key relationships and outlines the workings of this method.

The eye diagram, being the superposition of many contiguous pulses, hides statistical information of most key observable signal performance parameters listed above and repeated here for convenience:

- noise floor,
- min-max signal power levels,
- jitter,
- bit error rate (BER),
- signal to noise ratio (SNR), and
- pulse shape distortions (skew, kurtosis).

Assuming that a threshold level has been defined and also a timing point within a period on which the signal is sampled, two density distributions are observed, one for above the threshold and one for below it. The probability analysis shows that when the two probabilities overlap, the integral of overlapping area is the probability for error; that is the bit error rate (BER) (Figure 9.10).

Thus, based on the eye-diagram analysis, if the minimum (E_{min}) and the maximum (Emax), and also the eye openness (E_{eye}), are calculated, then, a parameter ρ, the

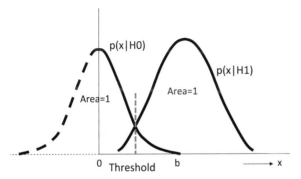

Figure 9.10.   Probability distribution functions for logic "one" and for logic "zero". The overlapping area of the two is the probability of error or BER.

standard deviations σ_1 and σ_0 and the means μ_1 and μ_0 of the corresponding distribution densities are calculated, then the following performance metrics are estimated:

$$SNR_{appr} = (\rho+1)/(\rho-1), \text{ where } \rho = E_{max}/E_{eye}$$

$$Q = (|\mu_1 - \mu_0|)/(|\sigma_1 - \sigma_0|),$$

$$BER = \frac{1}{2} erfc[Q/sqrt(2)], \text{ or}$$

$$BER = \frac{1}{2}\{1 - erf(V_p/[2\sigma_n \sqrt(2)])\}$$

Extinction ratio $R_{ext.} = \mu_1/\mu_0$

Noise factor $NF = Psi/kT_o\Delta f/SNR$.

Where *erfc* is the error function complimentary and *erf* the error function, the values of which are obtained from look-up tables. That is, the performance parameters BER, SNR, Q-factor, NF, Ext. ratio, and min-max power and more are all estimated by statistical sampling of the received signal. It is these performance parameters that characterize a channel's ID, which are stored in the receiver register.

Although the aforementioned method is based on eye-diagram sampling, the eye-diagram does not have to be monitored with sampling oscilloscopes. At the receiver, and behind the photodetector, the incoming pulses are sampled (this would correspond to the amplitude of the eye-diagram), the amplitude of sampled pulses is digitized and according to their amplitude range are stored in memory bins. It is these categorized values in memory bins that can be used to construct two virtual distribution functions, one for logic "zero" and one for logic "one," (Figure 9.11). From them, based on the aforementioned relationships, the performance parameters are estimated. Notice that the evaluated parameters provide a bulk indication of BER, SNR, etc., this method has no error detection or correction capability as error detection and correction codes (EDC) have. Nevertheless, the method provides an easy and very fast estimate of the signal performance that can be utilized in multiple ways, including security.

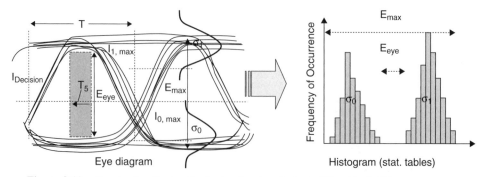

Figure 9.11.   Constructing the two distribution functions for logic "1" and logic "0".

9.4.4 Architecting the Performance Vector Circuit

Based on the description of the previous section, a performance estimation circuit may be constructed. The overall functionality is partitioned in two, one that performs sampling, analog to digital conversion and storage of digitized samples in corresponding bins. The other part performs calculations (μ, σ, Q, *BER*, *SNR*, *NF*, min-max values, etc.) and therefore a microcontroller is chosen; this allows for future software/firmware upgrades (Figure 9.12) [59, 60].

The converted optical signal to electrical is sampled and is compared to a threshold level. All samples are digitized and those that are greater than a threshold level are organized in bins in one part of a memory whereas those who are less than the threshold are organized in bins in another part of the memory. This organization is such that two virtual statistical histograms are formed from which the means and standard deviations are evaluated, the Q-factor, BER, SR, noise factor, power levels, and so on. At data rates of 2.5 to 10 Gbps, histograms are formed in microseconds and thus for all practical purposes, the performance vector is evaluated in real-time and in-service since real data are used. The estimated parameters are store in registers with a time stamp. A comparison of the current vector with previous ones yield the rate of change from which a possible fault/degradation or intrusion is detected.

Because the outlined method performs statistical sampling, and because optical bit rates are extremely high (>1 Gbps), a sufficient number of samples can be taken so that a performance vector can be completed within microseconds. However, in

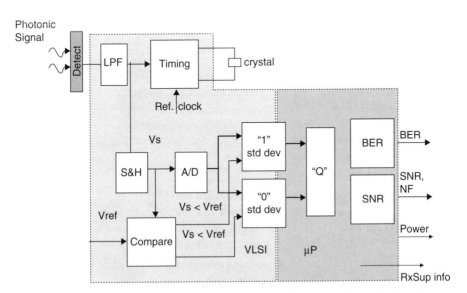

Figure 9.12. An integrated circuit is placed behind the photodetector to estimate the channel performance vector in real-time and without service disruption.

order to relax the switching specifications of the integrated circuit and the processing speed of the microcontroller, sampling can be one every 1000 incoming periods. This would slow down the calculations to yield a channel performance vector in milliseconds; in telecommunications this is still considered real-time.

Additional calculations are made to estimate the rate of change of a channel performance vector, for degradation/fault and also intrusion detection. To accomplish this, current values are compared with previous ones.

- Typically, if the ratio R = {current value}/{previous value} does not drop by more than 0.01 dB, it should not be of any concern, unless the degraded performance crosses the minimum acceptable threshold performance; in such cases, severe degradation may be viewed as failure.

- If the performance vector drops abruptly to the ground level, then the channel has lost connectivity. In this case, the performance vectors of other channels in the same fiber should be cross-correlated to identify whether there has been a single channel failure, a group channel failure or a fiber failure.

- If, however, the performance vector drops abruptly, yet remains above the acceptable threshold level, the likelihood of attack should be inferred and a countermeasure strategy should be followed.

9.4.5   Self-Protecting Optical Links

The following description considers a two-fiber DWDM link, one fiber per direction. In addition, the description assumes that the transmitting end of the link contains an association table with "data channel number" $\Leftrightarrow$ "wavelength number."

9.4.5.1   Intrusion Detection and Reaction Scenarios

The speed and efficiency of any countermeasure strategy depends on its ability to quickly monitor and detect fault and severe degradations, its ability to discriminate between faults/degradation and intrusions, and also on the speed to deploy a countermeasure strategy [61–63]. When Evan taps the physical medium, an abrupt degradation of the signal performance is detected by Bob's receivers, if they are equipped with the aforementioned circuit. Thus, upon detection of an intrusion, three possible actions may be taken:

- The network detects a malicious intrusion but does not react to it. However, it sends an alarm to the management unit and discontinues the transport of encrypted data. This may be called *detection with alarm and discontinuing service* (DADS). Instead, it may transport idle data.

- The network detects a malicious intrusion but does not react to it. However, it sends an alarm to the management unit and continues to transport encrypted data on the premise that the key is unbreakable by Evan. This may be called *detection with alarm and continuous service* (DACS).

- The network detects a malicious intrusion, sends an alarm to the management unit, and assumes that Evan is capable of breaking the code. In this case, the network deploys a countermeasure strategy, letting Evan believe that the encrypted data captured by him are good (that is, it sends encrypted decoy messages), although the true encrypted data have been moved to another channel or even medium following a simple algorithm. This may be called *detection with alarm and countermeasure intelligence* (DACI).

9.4.5.2   Self-Protecting with Countermeasure Intelligence Networks

The countermeasure strategy in the previous section is based on the following assumption:

- The optical network supports WDM technology.
- Optical links are full-duplex (single, dual-fiber or quad-fiber).
- Optical links include data channels and also an encrypted out-of-band (or optionally in-band) supervisory channel, or there may be redundancy, a working supervisory channel and a stand-by.
- Each port consists of a transmitter (laser), a receiver (photodetector and ancillary components), a performance vector monitoring circuitry (as already described), and a communication link to a management unit (on-board or out-of-board).
- Certain optical channels have been reserved for protection.

A version of the full-duplex link is shown in Figure 9.13.

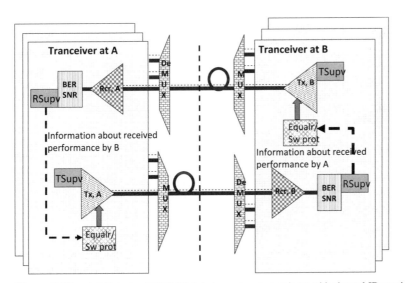

Figure 9.13.   A full-duplex DWDM link between two transceivers with channel ID monitoring.

Based on the link of Figure 10.20, the countermeasure protocol is as follows:

- The current performance vector is compared with previous ones. If there is no abrupt difference and the difference remains above a lower acceptable level, then the channel performance is considered stable. This level has been predetermined by channel parametric fluctuation analysis, which is beyond the scope of this description. In this case, the receiver sends to the source over the supervisory channel a ("NO DIFF") control message indicating no performance difference (Figure 9.14).

- If the difference is such that the channel is degrading and a degradation threshold is reached, then a control channel reassign request ("CHNL REASSIGN RQST") message is sent to the source over the supervisory channel. The latter searches for available protection channels from the pool of reserved ones and it responds with a wavelength (re)assignment control message (λ-REASSIGN") to the receiving end, while it moves traffic from the degraded to the newly reassigned wavelength. The receiving end assigns the newly assigned wavelength to the receiving port and starts monitoring the performance of the newly assigned channel; thus transmission continues with minimal or no traffic interruption. The same process is followed when a channel fails; in this case the performance difference is a step-function and the threshold is set much lower (close to the noise floor) than the degradation threshold.

- If after the reassignment to protection the degradation continues, the receiver sends another "CHNL REASSIGN RQST," and so on.

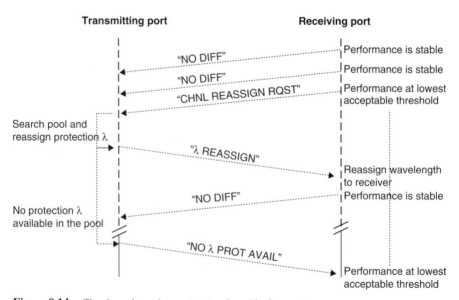

Figure 9.14.   The channel reassignment protocol used in the countermeasure strategy.

- If a second reassignment does not remediate the problem, then a link failure is declared by both the transmitting and receiving ends.
- If the transmitting end does not find any available wavelengths in the protection pool, then a "NO λ PROT AVAIL" control message is sent to the receiver, and in this case service continues even if it is degraded. In this case, the receiver continues monitoring the channel performance and periodically sends requests for new wavelength reassignment.

In this model, the degradation threshold is set somewhat higher than the minimum acceptable performance in order to compensate for protocol execution time during the channel reassignment. Thus, this *channel proactive reassignment algorithm* (CPRA) minimizes information (or packet) loss.

9.4.5.3 Simultaneous Multiple Channel Protection

Because each WDM port acts autonomously, multiple channel degradations may take place simultaneously. Clearly, the number of channels that can be protected can be as many as the number of protection wavelengths reserved at the source. However, considering that currently there are up to 320 wavelengths in the C and L bands, then one of the well-known protection strategies (1 + 1, M : N) may be adopted. In fact, by combining this strategy with the channel proactive reassignment algorithm (CPRA), a fast switch to protection is achieved.

Besides degradations and failures, the difference of the channel performance vector may reveal malicious attacks on more than a single channel. In such cases, the algorithm for channel protection is the same with the addition of the following.

The receiver sends to the source a request for channel reassignment with continued service over the existing channel ("CHNL REASSIGN RQST+"). Based on this, service over the compromised channel continues but data is neither sensitive nor proprietary, whereas sensitive encrypted data passes over the newly assigned channel that only the source and destination know about.

In this scenario, if channels keep being compromised, then the complete link may be declared compromised and under attack and traffic may be routed to another fiber, following a link protection strategy 1 : 1 or M : N.

9.4.5.4 Simultaneous Multiple Channel Equalization

When channel degradation is detected but the degradation has not reached the lowest acceptable performance, it may be necessary to perform channel equalization at the transmitter side. In this case, the receiver requests a channel equalization ("CHNL EQAUL RQST"). The transmitter receives this message and if equalization is supported, it adjusts the transmitting power either in small increments or in one large step.

Because this model is based on a closed loop system, eventually the channel will be equalized and the receiver will stop sending the request.

Based on this method, because each receiver monitors its own received channel, it is possible that multiple channels require equalization, some more than others. In this case, receiver sends individual ("CHNL EQAUL RQST") messages, and each transmitter performs adjustments independently. Thus a simultaneous multi-channel equalization is achieved potentially in less than milliseconds; current practice takes many minutes as equalization is accomplished channel at a time.

9.4.5.5 Intrusion Counter-Attacking Networks

The aforementioned countermeasure strategies based on the performance vector circuit, give rise to a brand-new area in network security, which is beyond the self-defending networks described above: *Counter-Attacking Networks* (CAN).

Counter-attacking networks require a little more intelligent and sleuth. The assumption is that Evan is very sophisticated having to his disposal computer-based equipment and he attempts to copy data, inject data and break codes.

When intrusion detection takes place, then the receiver has notified the transmitter about it, and the transmitter does two things. It sends over a secure supervisory channel the channel reassignment which has been modified so that service continues uninterrupted, and sends decoy messages over the attacked channel, which Evan copies but the receiver of the link kills. The reason for the latter is that the decoy messages contain malware (a virus, a Trojan horse, or a worm), which when copied by Evan nests in his computer and creates havoc to him. Because it is killed at the receiver, this malware does not propagate the network. Thus, the network in its own way counter-attacks the intruder.

In this scenario, the malware injected in Evan's computer can be more sophisticated. It may lead Evan to wrong keys, or to decipher a wrong text, or to believe that he has been successful where he is not, and so on.

In summary, it is clear that a secure network may employ self-defending and counter-attacking strategies against overt attacks. The aforementioned countermeasure strategies, self-defending and counter-attacking networks are two of several possible scenarios in network security at the PHY layer. Research continues into other scenarios, including aggressive networks [64].

9.5 A WDM METHOD APPLICABLE TO LINK SECURITY

Wavelength Division Multiplexing (WDM) is an optical and photonic technology which multiplexes many optical signals, each at a different wavelength, and couples them onto a single optical fiber [1]. Optical channels are within the fiber low attenuation spectral band (from approximately 1300 to 1620 nm) and are ITU-T defined in two different ways.

- ITU-T G.694.1 defines a dense grid of 160 optical channels separated by 50 Ghz (or 320 separated by 25 Ghz) over the spectrum 1510–1620 nm (C + L bands) using standard single mode fiber, and hence dense wavelength division multiplexing (DWDM) [65].

- ITU-T G.694.2 defines a coarse optical channel grid, in which channels are separated by 20 nm over the spectrum 1261–1621 nm using water-free fiber, hence coarse wavelength division multiplexing (CWDM) [66].

DWDM is applicable to long-haul and backbone network applications that demand a plethora of channels in order to transport an aggregate traffic of several Tbps.

CWDM is applicable to short-haul and access, such as Metro LAN and fiber to the premises (FTTP), to transport aggregate traffic up to 500 Gbps at distances less than 40 km.

Despite the differences, both WDM technologies transport impressive bandwidths when they are compared with other bandwidth transporting technologies and media (wireless and wired).

WDM technology has already been established in optical communication networks as well as in computer grid networks. However, when massive amounts of information are transported there are three issues that stand out: delivery of ultrahigh bandwidth, cost-effective security of information (data, application), and security of the network (transporting link and path). Security of information is addressed with encryption techniques and algorithms, and of the link with countermeasure strategies as already described. This section describes a WDM link security method and algorithm, which may be combined with the performance vector to render Evan's channel eavesdropping virtually impossible.

9.5.1 The Wavelength-Bus

In DWDM communication networks, information from various tributary sources neither arrives at the same rate nor conforms to the same protocol. The next generation optical network defines dissimilar traffic (synchronous and asynchronous) encapsulated in a single synchronous payload transported over a single WDM channel [REF12-REF15]. For practical reasons, all WDM channels in a single fiber are modulated at the same bit rate producing optical serial bit streams, each stream at different wavelength. For simplicity of description, consider a group of eight streams, S_1 to S_8, and 8-bit octets or bytes although 10-bit or more may be used (Figure 9.15).

S_1: a_{10} a_{11} a_{12} a_{13} a_{14} a_{15} a_{16} a_{17} b_{10} b_{11} ...
S_2: k_{20} k_{21} k_{22} k_{23} k_{24} k_{25} k_{26} k_{27} ... a_{20} a_{21}
S_3: m_{30} m_{31} m_{32} m_{33} m_{34} m_{35} m_{36} m_{37} ... a_{31}
  ... : ....
S_8: p_{80} p_{81} p_{82} p_{83} p_{84} p_{85} p_{86} p_{87} ... a_{80} a_{81}

Figure 9.15. DWDM multi-channel data organization (the first index denotes stream number {1:8} and the second denotes bit number {0:7}).

Rail λ_1: a_{10} k_{10} m_{30} ... p_{80} b_{10} ...
Rail λ_2: a_{11} k_{11} m_{31} ... p_{81} b_{11} ...
Rail λ_3: a_{12} k_{12} m_{32} ... p_{82} b_{12} ...
Rail λ_4: a_{13} k_{13} m_{33} ... p_{83} b_{13} ...
Rail λ_5: a_{14} k_{14} m_{34} ... p_{84} b_{14} ...
Rail λ_6: a_{15} k_{15} m_{35} ... p_{85} b_{15} ...
Rail λ_7: a_{16} k_{16} m_{36} ... p_{86} b_{16} ...
Rail λ_8: a_{17} k_{17} m_{37} ... p_{87} b_{17} ...

Figure 9.16.   The wavelength-bus.

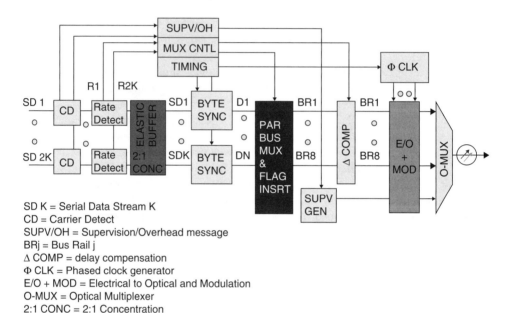

SD K = Serial Data Stream K
CD = Carrier Detect
SUPV/OH = Supervision/Overhead message
BRj = Bus Rail j
Δ COMP = delay compensation
Φ CLK = Phased clock generator
E/O + MOD = Electrical to Optical and Modulation
O-MUX = Optical Multiplexer
2:1 CONC = 2:1 Concentration

Figure 9.17.   Block diagram of the WDM l-bus multiplexer.

Consider that bytes or octets in each serial channel are converted to parallel, and that bytes from several parallel data streams are multiplexed based on a multiplexing algorithm that accommodates data streams of different data rate. For 8-bit bytes this constructs an 8-bit wide bus which in the optical layer consists of 8 different wavelengths or rails, where all eight rails are transmitted at the same bit-rate. This is called a *WDM wavelength-bus* or a *λ-bus* [67–70]. The end result of eight serial channels converted to an eight-rail λ-bus is shown in Figure 9.16.

A fundamental distinction is important: Figure 9.15 shows many sources from various tributaries (this is important for this link security algorithm) as opposed to a computer single source parallel bus [16], which represents a simplistic case of this method.

Similarly, Figure 9.17 illustrates the multiplex section of tributaries and a buffer that multiplexes parallel bytes according to a user-defined algorithm. The buffer allows for a 2:1 aggregation (the equivalent of 2:1 concentration) as well as

Table 9.3. Possible permutation according to buffer depth

Buffer depth (bytes)	Possible permutations
1	$2^8 = 256$
2	$2^{16} = 65,536$
3	$2^{24} = 16,777,216$
4	$2^{32} \sim 4.3E9$
5	$2^{40} \sim 1E12$
6	$2^{48} \sim 2.8E14$
7	$2^{56} \sim 7.2E16$
8	$2^{64} \sim 1.8E19$

smoothing-out the random arrival of bursty packets, whereas the electrical to optical conversion and modulation is at the output physical layer. A similar circuit in reverse provides the demultiplexing function at the receiver.

Based on this, at the interleaver and depending on buffer size, the possible bit permutations for an 8-rail wavelength bus are, Table 9.3:

9.5.2 Bandwidth Capacity and Bandwidth Elasticity of the λ-Bus

The bandwidth capacity C of the λ-bus is,

$$C = N \times D \, (\text{GHz})$$

where D is the bit rate per rail and N the number of rails used on the λ-bus.

The rule to observe is:

$$C \leq \sum n_j C_j, \text{ over all } j$$

where n_j is a positive real number between 0+ and N, and C_j is the data rate of each serial tributary prior to been multiplexed onto the λ-bus. For example, for an 8-rail λ-bus, each rail at 10 Gbps, then the λ-bus capacity is 80 Gbps.

The bandwidth elasticity of the λ-bus is demonstrated with three examples:

EXAMPLE 1: *Consider dissimilar traffic WDM multiplexed as follows:*

5 channels at 10 Gbps each,
2 channels at 2.5 Gbps each, and
4 1GbE channels.

If for each source a different optical channel is used, then it would require 11 channels, since $i = 11$. However, using the λ-bus, the total bandwidth fits in the λ-bus capacity and thus 8 channels are used.

EXAMPLE 2:   *Consider dissimilar traffic WDM multiplexed as follows:*

1 channel at 40 Gbps,
2 channels at 10 Gbps each,
4 channels at 2.5 Gbps (here $i = 7$).

If the bit rate of 40 Gbps is not supported by the link, this would have been impossible. However, the λ-bus makes it impossible as the sum of the bandwidth of 70 Gbps fits within the 80 Gbps λ-bus capacity.

EXAMPLE 3:   *Consider dissimilar traffic WDM multiplexed as follows:*

1 channel at 80 Gbps.

This case would have been impossible since the currently maximum bit rate commercially available is 40 Gbps. However, the λ-bus makes it impossible.

An additional important benefit of the λ-bus is following: 40 Gbps channels are uneconomical as they can be used in short links without serious impairments due to non-linear phenomena (polarization mode dispersion or PMD, environmental temperature variations, and others); however, transporting 40 Gbps over the λ-bus (each rail at 5 Gbps) the fiber reach is in the order of 60–80 km without amplification. Thus, the λ-bus is flexible enough to accommodate elastic bandwidth, reduces hardware and software resources at the physical layer, and reduces system design complexity and cost [71].

9.5.3 Security Aspects with the λ-Bus Based on a Multi-Strand DNA Model

This section makes the assumption that information from each tributary or source is already encrypted by the end-user according to an established encryption or scrambling method. Subsequent to this, the tributaries are multiplexed to a λ-bus, as described. Thus, each rail on the λ-bus consists of bits that belong to a different cipher text, ciphered with different encryption keys and different encryption algorithms. Moreover, the actual wavelengths (or channels) used to construct the λ-bus are not known to Evan but only to transmitting–receiving ends, which select 8 out 160 (if an 8-rail bus is constructed); these channels may have been predetermined and dynamically reassigned from time to time. Thus, when tapping the fiber and attempting to copy a channel, the channel provides no meaningful clues to Evan.

However, if Evan is highly sophisticated with high-tech equipment and supercomputers, one has to enhance the security of the optical link, from the transmitting point, called point-A or "Adam," to the receiving point, called point-B or "Bob."

This section describes two securing methods, the "multiplexer random scheduler" and the "temporal λ-bus randomizer," and also a method that combines both

[72, 73]. In fact, some of the security aspects of the bus emulate a multi-strand DNA model [74].

9.5.3.1 Multiplex Random Scheduler

Consider that the multiplexer receives a variety of payloads at the same or different data rates.

If the rates are the same, then for an eight rail λ-bus {A, B, C, ... , H} an interleaver randomly multiplexes bytes from each, based on a random algorithm, producing eight randomized streams (for example, one of the streams may be ABBACDDFEFFAC ...) each one mapped onto one of the rails of the λ-bus. The set of eight wavelengths that comprise the wavelength-bus and the random algorithm is known only to the transmitter and to the receiver where the streams are de-interleaved.

Now, assume that the interleaving algorithm does not remain the same but also varies at random intervals, according to a chaotic algorithm. Information about the key that defines the interleaving sequence and the period may be predetermined, may be generated by a predefined secret method, or may be transmitted to the receiver over a secure supervisory channel. The method of generation and communication of such keys follows standard cryptographic methods, as elaborated in this book.

If the rates are not the same, then the interleaver multiplexes more bytes from the tributary with higher rate and fewer from tributaries with lower rate, while adding recognizable (by the receiver) idle bytes to fill the l-bus capacity C (filling in with recognizable idle patterns is common practice in many modern data protocols).

9.5.3.2 Temporal Wavelength-Bus Randomizer

Consider that the set of parallel data streams $D(t)$, prior to multiplexing onto the λ-bus undergoes a random bit-ordering rotational transformation $T(x, t)$, producing a new data stream $D(x(t), t)$ (Figure 9.18):

$$D(x(t), t) = T(x, t) \times D(t)$$

One may envision that the random bit-ordering rotational transformation consists of two aligned wheels, each divided in eight 45° sectors {S0A to S7A} and {S0B to S7B}. In this case, each sector reads bits from a rail of the 8-bit λ-bus while still in the electrical regime.

Wheel A receives the data set $D(t)$, performs the rotational transformation $T(x, t)$ by rotating at random increments k45°, where k is a random integer between −8 and +8, and transfers the product to wheel B which is fed to the electrical-to-optical (laser + modulator) device. Thus, the bit order on the wavelength bus is temporally randomized. This mechanism is called temporal wavelength bus randomizer.

As in the previous method, this method also may employ a chaotic algorithm so that the random bit-ordering does not repeat but changes at random intervals.

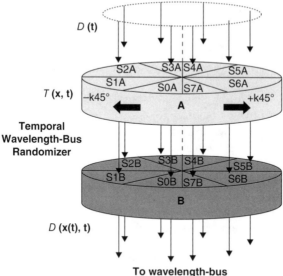

Figure 9.18.   The analog of temporal wavelength-bus randomizer.

9.5.3.3   The Hybrid Method

It is clear so far that data from several sources on a λ-bus has been encrypted in several ways:

- at the information level each end-user uses specific cipher keys and an encryption algorithm, which is different from one end-user to another.
- at the physical layer data is encapsulated in a virtual channel according to the transporting protocol.
- by the random interleaving scheduler that maps data from multiple sources onto a λ-bus,
- by the randomizing process of the λ-bus, and
- by the selection (perhaps random) of optical channels or wavelengths to make up the λ-bus, which unknown to an eavesdropper.

Based on this, for Evan it is virtually impossible to isolate a particular client stream, monitor a multiplicity of keys and de-encrypt a cipher text. The receiver, for routing purposes, is able to de-interleave the tributaries but is not able to decrypt client data; only the end-user can do this. Figure 9.19. shows the encryption hierarchy from client to the WDM physical link layer.

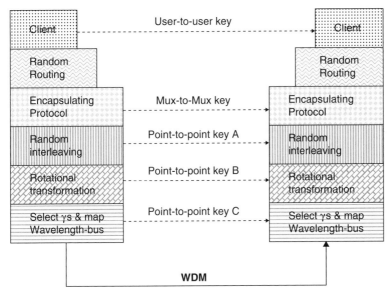

Figure 9.19.   Hierarchical data encryption based on multiple cipher keys with chaotic behavior transported to corresponding peer level across the network.

9.6 FREE SPACE OPTICAL NETWORKS

As soon as laser technology was established in fiber-optic communications, it was realized that modulated laser beams will be applicable to applications without fiber but as in a line-of-sight point to point communication. There have been two cases in this area: one is applicable to inter-satellite communications [75, 76] and the other is in terrestrial applications, called free space optical communication [77]. This section investigates the terrestrial application and its security aspects.

9.6.1 FSO Technology Overview

The DWDM fiber-optic communications technology being able to transport in the backbone network more than 1 Tbps aggregate bandwidth per fiber has challenged the copper-wire access network; this has been dubbed as *first/last mile bottleneck*.

The passive optical fiber network (PON) known as fiber to the premises (FTTP), and more generally FFTx, combined with the deployment of new broad-band mobile wireless services (WiMax and IMS) is expected to remedy the bottleneck problem.

However, FTTP fiber installation in many cases and particularly in rough terrains is tedious and requires long planning which many clients cannot wait for. In such cases, an alternate optical network solution that can be quickly deployed and offer very high bandwidth may be used. This solution is known as free space optical (FSO) technology.

FSO technology uses two laser-photodetector transceivers to establish a line of sight duplex point-to-point link. Typical FSO links operate between 10 Mbps to 2.5 Gbps in the range 800–1550 nm with optical transmit power −8 dBm that can reach an operating distance up to 3 km.

FSO transceivers are mounted on top of or at a window of tall buildings. In either case, FSO transceivers require accurate alignment because the laser beam is relatively thin and the receiver aperture has a small diameter. After the initial alignment, automatic tracking maintains alignment during strong winds that sway tall buildings by few meters. Typically, an FSO link can be established and become operable within a day or less with relatively inexpensive technology to meet immediate communication needs in the short-term.

However, FSO technology has its own shortcomings. As the laser beam propagates in the atmosphere it suffers attenuation, scattering, scintillation, and angular dispersion by various airborne particles [78]. Here, for completeness, is a brief synopsis of these impairments.

Attenuation: The density of particles and molecules in the beam's path define the limiting conditions of link distance and operational data rate that meet the expected link performance. Two factors play a key role, laser wavelength, and the particle chemical and physical consistency on the beam's path; the longer the wavelength, the lower the atmospheric attenuation. Dense snowflakes and fog cause severe attenuation of the laser beam. Thick fog causes 50 dB/km attenuation that severely degrades the link performance, whereas medium fog causes about 30 dB/km. Forward error correction codes (FEC) may also be used to extend the link distance by 20–30% or to improve the data rate.

Another factor that affects attenuation is its dependence on wavelength, so that the 1550 nm wavelength is preferred over the 800 nm. However, VCSEL laser technology at 800 nm is more economical and for moderate data rates including 1 GbE many FSO systems operate at this wavelength with optical transmitted power about 10 mW; optical power varies among manufactures and network applications.

Scattering: Scattering is the contribution of three mechanisms, Rayleigh, Mie and Geometric, which depend solely on particle size and particle density.

Scintillation: Atmospheric turbulence due to temperature and pressure variations causes additional attenuation on the laser beam; this is known as scintillation noise and is manifested as a speckled pattern. However, scintillation is responsible for few dB and is a small attenuation contributor compared with fog [79].

If in certain geographical areas thick fog is frequent, then an RF back-up link is used to protect the FSO link. Typical RF links operate at the 60 Ghz license-free band (in the US and Canada) and are not attenuated as much by snow and fog as by heavy rainfall.

Alignment: In point-to-point FSO links, the transmitter-receiver or link alignment is extremely important. The laser beam is initially very thin and diverges slightly; beam divergence of 1.3 mrad will cover a circle of approximately 6 m at 5 Km. Beam divergence helps to initially align the link and to also compensate for

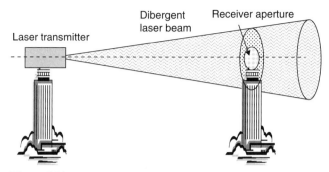

Figure 9.20. The divergent FSO beam facilitates pointing, acquisition and tracking. (shown half-duplex and not in scale).

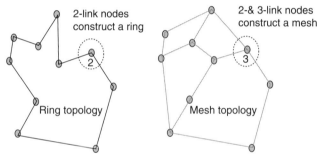

Figure 9.21. FSO nodes with two or three transceivers are interconnected to construct a ring or a mesh network.

building sway due to environmental conditions (wind, temperature). Divergence is also seen as negative gain since the aperture of the receiver is a fraction of the beam cross-section at the receiver and thus not all optical power is utilized (Figure 9.20).

9.6.2 FSO Mesh Topology

The limitations of a point-to-point FSO link are obvious. In applications where an FSO communications network needs to provide wide area coverage, multiple FSO transceivers may construct a mesh topology (Figure 9.21). This is a relatively new addition to the FSO networking technology, that was [80], and may be implemented using one of two methods.

- Multiple FSO transceivers are stacked in a housing header to provide necessary add-drop, routing and switching functionality using off-the-shelf parts that conform to popular standard protocols (SONET/SDH, GbE).

- Multiple FSO transceivers are stacked in the same housing in which an integrated circuit also provides (via provisioning) the necessary add-drop, routing and switching functions [81, 82]. This method simplifies hardware and node maintenance, and lowers power consumption and cost.

According to Figure 9.21, FSO mesh nodes consist of multiple transceivers. Depending on network complexity, nodes may have two, three or more transceivers. However, it has been estimated that the optimum number of transceivers per node is up to three, considering complexity, alignment, sway modes, switching options, topology, terrain, and cost. Three transceivers per node construct a honeycomb-like mesh network, as is the case in cellular mobile technology. This configuration exhibits the following advantages [83, 84]:

- larger traffic capacity
- multiple locations interconnected
- better routeability
- better network performance
- better traffic management (balancing, grooming, and congestion avoidance)
- simpler and better link protection
- better service protection
- better fault management
- better fault avoidance (for single and cluster faults)
- enhanced security features

9.6.3 FSO Protection Strategies and Fault Avoidance

The mesh topology may consider either half-duplex or full-duplex links. Although full-duplex offers superior protection against node or link failure [85] and utilizes the full network capacity, half duplex simplifies node complexity while offering adequate protection and acceptable service. In either case, when a faulty node is detected, traffic is rerouted over another path according to well-known rerouting algorithms.

When a link fault or a severe degradation is detected, the link and node needs to be tested. This is accomplished by performing traffic loop-back tests. That is, the received data stream is looped back by provisioning the node switching fabric; this is possible with maintenance and controlled packets (according to SONET/SDH and GbE protocols).

FSO mesh networks employ fault and performance monitoring mechanisms, fault management and protection strategies similar to landline networks. Link performance is monitored by estimating BER and received power using error detection and correction codes (EDC), and possibly statistical estimation methods.

9.6.4 FSO Security

The FSO mesh network has several inherent security features [86]:

- the laser beam is invisible to the human eye;
- the laser beam is relatively very narrow;
- the beam is at a height (as measured from the ground) that is not easily accessible. Therefore tapping the link is not easy, at the exception of the transceiver housing, which is typically on a rooftop.

However, there are also few shortcomings:

- The beam is slightly divergent and some optical power is transmitted beyond the receiver, which may be exploited for eavesdropping (Figure 9.22). This is avoided by either properly positioning the receiver or placing a panel behind it to block the overspill.
- FSO networks deployed in dense building areas, the beam may pass near a physical object where a transparent panel may deflect part of the beam and redirect it to another photodetector for eavesdropping.
- If the receiver is behind a window, the glass panel may deflect part of the beam and be redirected to a location where the eavesdropping photodetector could be placed. However, careful surveying and inspecting can avoid such eavesdropping cases.

In general, FSO transceivers are installed on rooftops of tall building and thus the transceivers are not easily accessible, the electronics may be enclosed in secure enclosures with sensors and alarms and the interconnecting cables may be protected by tamper-resistant conduits. Finally, because FSO networks are targeted for private network applications, a private encryption protocol with individualized cipher keys may be used over each link, which could frequently be changed according to a private schedule.

Although FSO links exhibit better security features, the communication equipment are have similar vulnerabilities to those of any network. Particularly, if access is wireless, then access is also vulnerable as already discussed, in which case authentication protocols, cryptographic algorithms, and cipher keys play an important role in information security.

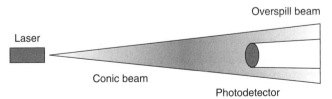

Figure 9.22.   The divergent FSO beam overspill the receiver presenting an opportunity for eavesdropping, if it is not blocked.

9.7 SECURITY TESTING

In general, the strength of security depends on the weakness of the intruder. In cryptography, codes that were believed to be unbreakable have been broken, and newer codes challenge crypto-analysts to break them, if they can, and over time, there have been announcements of which code was broken. Therefore, network security methods and particularly math-intensive cryptographic algorithms need testing prior to deployment.

Testing cryptographic tools is more of an art than a systematic methodology with conclusive scientific results. However, cryptography is math intensive and follows complex algorithms. Therefore, testing can be focused on two areas, the math part of the cryptographic code and the algorithmic part of it.

- With respect to the math part, there are specific components that can be tested. These tests use statistics and probabilities to provide not so much a 100% good/bad result but a probability of how difficult it is to solve a specific problem that leads to breaking the code. Such a component in cryptography is the random number generator (RNG) or the pseudo-random number generator (PRNG).

- With respect to the algorithmic part, an algorithm is provided by several tools: a flow graph that provides the logical sequential process, which based on input parameters and decision follows a specific path on the graph; a state diagram—based on stimuli and conditions at the current state—decides which is the next state to move to; and by recursive equations. Any of these methods help develop software or firmware that the computing machine will execute.

9.7.1 RNG and PRNG Testing

In general, random number generators (RNG) and pseudo-random number generators (PRNG) produce a sequential binary stream of logic "zeros" and "ones," which may be random over its full length or random over blocks of it. In either case, the generated stream may be subdivided into blocks or subsequences of random numbers. The length of the random number that is generated depends on the RNG or PRNG polynomial and on the seed or initial value (IV).

In cryptography, RNGs and PRNGs are pivotal in the generation of cipher keys. RNGs should conform to both forward unpredictability and backward unpredictability (see Chapter 2). That is, the RNG produces a stream such that the next bit is unpredictable even if the seed is unknown, and that the seed cannot be derived from the generated random stream [87]. Therefore, it is important that the uniqueness and randomness of the stream is tested based on well-defined rules.

Occasionally, an output may appear random, such as electronic noise, which, however, after statistical and Fourier analysis, shows that it is the superposition of random noise (such as Thermal noise) and a periodic signal, such as 60/50 Hz or another periodic phenomenon that influences randomness.

To test randomness, there are two fundamentally different testing methods:

- Statistical tests attempt to estimate probabilistically the true randomness of a sequence.
  - There are 16 statistical probabilistic tests, but no one test provides a "complete and conclusive" answer on the randomness of a stream. Such methods test whether a specific pattern of length n within a stream is truly random. Among them are the "test for the longest-run-of-ones in a block," the "frequency test within a block," the "discrete Fourier transform (spectral) test," the "linear complexity test," the "serial test," the "approximate entropy test," the "random excursion test," and so on (for a complete discussion of all 16 tests, (see [87]).
  - Among the many parameters provided by these tests, a measure of the disorder or randomness of a stream is given by the entropy of uncertainty of a random variable X with probabilities $p_1, p_2, \ldots, p_n$, which is defined as $H(X) = -\Sigma p_i \log p$, where the sum is from $i = 1$ to n. For a random variable, with $p(0) = p$ and $p(1) = 1 - p$, the entropy is expressed as $H(X) = -[p \log p + (1 - p)\log(1 - p)]$ which is a concave function with a maximum value of unity at $p = 1/2$.
  - Similarly, the "randomness" quality of a sequence under test is given by a P-value, also called the *tail probability*. The P-value is defined as "the probability (under the null hypothesis of randomness) that the chosen test statistics will assume values that are equal to or worse than the observed test statistic value when considering the *null hypothesis*" (see [87]).
  - In sequence testing, the *null hypothesis* H_0 supports the sequence under test is random whereas an alternative hypothesis H_a supports that the sequence is not random.
- The brute force approach tries to identify repetitive random patterns of different lengths in the generated stream by the variable window scanning method.

Either of the two methods starts with certain assumptions, with respect to sequence randomness, uniformity, scalability and consistency.

- Uniformity means that the next bit in the sequence has equal probability of being a "zero" or a "one;" in communications this is expressed as the probability of generating a "zero" or a "one" is equal, that is $\frac{1}{2}$. Thus, in a stream of n bits, $n/2$ are "ones" and $n/2$ are "zeroes."
- Scalability means that any test for randomness applicable to a sequence is also applicable to any size block of the sequence.
- Consistency means that the RNG always generates the same output for a given seed (IV). For a different seed, the output is different but it is always the same for the same seed.

In conclusion, the strategy for testing a RNG follows the following logical steps (reference NIST indicates that sample data may also be obtained from George Marsaglia's *Random Number CDROM*, at http://stat.fsu.edu/pub/diehard/cdrom/):

1. Select a binary sequence generator (hardware or software-based) for evaluation. Examples and methods may be found in ANSI X9.17 (Appendix C), and in FIPS 186 (Appendix 3).

2. With the selected binary generator and for a fixed sequence of length n, construct a set of m binary sequences.

3. Invoke the NIST Statistical Test Suite using the desired sequence length and the m binary sequences produced in step two. Select the statistical tests and relevant input parameters (e.g., block length) to be applied.

4. An output file is generated by the test suite with test statistics results and *P-values* for each statistical test. Based on the *P-values*, a conclusion regarding the quality of the sequences can be made.

5. Make a pass/fail assessment.

9.7.2 Cryptographic Algorithm Testing

Flow graphs and state machines are powerful tools in the design of algorithmic decision-making processes. Given the complexity of these tools in cryptography, the testing objectives are to identify in them traps, endless loops, and dead-ends.

Typically, a trap is a condition where the state is "stack" and does not move to the next logical state regardless of what the stimulus is. A loop is when the machine moves from a state to the next and then back again and it does not progress to the "finish" state; that is, the state machine is "stack: between two (or more) states. A dead-end is also a "stack at" condition but at an end of the state machine, which does not return to the "start" state.

Testing of state machines and flow graphs are well known in CAD (computer aided design) tools. Because these tests use binary logic, as is used in computer-based cryptography, CAD tools also can test cryptographic algorithms. For this, a sequence of binary test vectors is generated, either based on thorough knowledge of the state machine or flow graph functionality, or by generating binary sequences at random.

The state machine or flow graph is converted in a descriptive language that the CAD tool understands. Then, the generated vectors stimulate the inputs of the state machine sequentially and the computer tracks the states of its node and link in the state machine. Running the vectors through, the computer identifies "stack at" conditions, which are caused either by incomplete vector development or by flaws in the state machine. Design inspection of both resolves the malfunctioning and eventually a robust design and comprehensive test is developed. These test vectors are later used to identify future failures of the developed product.

State machines (SM) and flow graphs (FG) may be static, or dynamic.

- Static SMs remain at a state until the next stimulus arrives.

- Dynamic SMs remain at a state for a pre-assigned period and then they move to the next state autonomously. If a stimulus arrives prior to the pre-assigned period, then the state moves to the next state based on the value of the stimulus that arrived. Dynamic SMs and FGs are also known as time-dependent SMs and FGs.

Testing dynamic SMs and FGs is more tedious than static ones, particularly if the defined period at each state is variable.

- When testing static SMs and FGs, a uniform period is defined, which typically is the clock period of the tester.
- When testing dynamic SMs and FGs, it is typical to define the pre-assigned period at each state in integer multiples of the clock period; however, testing becomes more complicated if the pre-assigned period is analog (not an integer multiple of the clock period).

9.7.3 Cryptanalysis as a Security Testing Tool

For completeness of this chapter this section briefly reviews the reverse of cryptography—cryptanalysis. It is not considered a treatment on this subject.

To contrast the two, one may think of cryptography as similar to designing a circuit that performs a specific secret function, and then enclose the complete circuit in a sealed black box, A. Data enter box A, and different data exit box A, and the output is sent to another black box, B, which is designed to undo what box A did, either executing the process in reverse, or using a mathematically inter-related process.

Now, think of cryptanalysis as looking at the black box A and trying to identify what the exact process is within the box, or at least identify the key functions that are needed so that ciphered data on the link, between box A and B, can be deciphered. The rule is that one cannot use x-rays to get a glimpse inside the box but devise tests and social tricks that may create some knowledge on the statistical behavior of the cipher that may help cracking it.

Often, cryptologists challenge cryptanalysts on breaking a newly developed cipher. As an example, consider the $100 RSA-129 challenge that was made in 1977; a cryptographic algorithm with a 129-digit key produced a cipher text. The claim was that if one tries to break the cipher text that was produced with this long key and read it, it would take 40 quadrillion years (or several million years). Unfortunately for this code, it did not take so long (read the eloquent article "*Wisecrackers*" by columnist Steven Levy, steven@echonyc.com, published in The Condé Nast Publications Inc., Copyright © 1993–2004).

In all, the job of the cryptanalyst is more difficult than that of the cryptographer, and although cryptanalysis may seem to be an intrusion to data privacy, which it is if unauthorized, cryptanalysis is also a useful tool to test cryptographic methods. As a result, several cryptanalytic methods and tools have been developed, such as classical cryptanalysis, differential cryptanalysis, and others, the description of which is beyond the purpose of this book.

However, for completeness here follows a description of a basic statistical cryptanalytic method that is based on the frequency of occurrence of letters in words. When a single letter is considered for the frequency of occurrence in a text, this is called a *unigraph*, when two adjacent letters are considered, it is called a *digraph*, and when three a *trigraph*.

For example, the letters *E, T, N, R, O, A, I* and *S* appear in 67% of English words, and the vowels *A, E, I, O* and *U* appear approximately up to 40% in a relatively long English text. Similarly, common digraphs are *TH, HE, EN, RE,* and *ER,* and common trigraphs *THE, ING, CON, ENT* and *ERE.* For short texts, the frequency of occurrence differs.

Having a cipher, a frequency count is made to identify the most and the least common letters in it. From this, some statistics are formed from which certain deductions can be made. Based on this, if the cipher used the transposition method, the frequency of occurrence of letters in the original text result in the same frequency of occurrence of the transposed letters; this provides a clue towards breaking the cipher.

Simple ciphers produce a frequency distribution of letters with sharp peaks and valleys; these are called *rough* ciphers. More complex ciphers tend to flatten the frequency distribution; the flatter the distribution, the better the cipher.

In statistical cryptanalysis, several factors help in breaking the cipher. Among them are:

- language used in the text,
- knowledge of the type of text (technical, legal, etc.),
- punctuations used,
- tendency to use long or short words,
- richness of vocabulary,
- text padding,
- obtaining a set of ciphertexts coded with the same method,
- obtaining the ciphertext and the plaintext (from which the cipher key may be determined),
- obtaining subsequent ciphertexts to study trends or on cipher development, also known as adaptive cryptanalysis.
- if a specific word is repeated (the hieroglyphic and hieratic scripts on the Rosetta stone were decoded from the simple word <Cleopatra> that was written in these two undecoded languages and in Greek),
- and, so on.

Besides statistical cryptanalysis, external information may also help in cracking the cipher. For example, when Evan succeeds in mimicing a source, he may pass through the system test texts that help identify the workings of the cipher algorithm and the keys that are used. Thus, in addition to strong cryptography for information privacy, strong authentication and firewalls, network security with countermeasures and aggressive tactics—such as counter-attacks—is equally important for winning the cyber war.

REFERENCES

1. ITU-T Recommendation G.652, *Characteristics of a single-mode optical fiber cable*, 1997.
2. S.V. KARTALOPOULOS, *Introduction to DWDM Technology: Data in a Rainbow*, (Published in US and in India) Wiley/IEEE Press, 2000.
3. S.V. KARTALOPOULOS, *DWDM: Networks, Devices and Technology*, IEEE/Wiley, 2003.
4. S.V. KARTALOPOULOS, "DWDM: Shaping the future Communications Network", *IEEE Potentials Magazine*, October/November 2004, pp. 16–19.
5. S.V. KARTALOPOULOS, "Elastic Bandwidth", *IEEE Circuits and Devices Mag.*, vol. 18. no. 1, pp. 8–13, Jan. 2002.
6. S.V. KARTALOPOULOS, "What is DWDM?" SPIE Electro-Optic News, Nov. 2000, also available in http://www.spie.org/web/oer/november/nov00/wdm.htm.
7. S.V. KARTALOPOULOS, *Optical Wavelength Division Multiplexing (WDM) Networks and Technology*, IEEE, an audiovisual short course, IEEE ExpertNow series.
8. S.V. KARTALOPOULOS, *DWDM Technology*, IEEE Communications Society, Tutorials-Now series.
9. ITU-T recommendation G.694.2, "*Spectral grids for WDM applications: CWDM Wavelength grid*", (6/2002).
10. A. SIERRA and S.V. KARTALOPOULOS, "Evaluation of Two Prevalent EPON Networks Using Simulation Methods", Proceedings of the Advanced International Conference on Telecommunications 2006 (AICT'06), 2/19–22/06, session AICT-3, Guadeloupe, French Caribbean, on CD-ROM, product: E2522, ISBN: 0-7695-2522-9, Library of Congress: 2005937760.
11. S.V. KARTALOPOULOS, "Next Generation Hierarchical CWDM/TDM-PON network with Scalable Bandwidth Deliverability to the Premises", *Optical Systems and Networks*, vol. 2, 2005, pp. 164–175.
12. S.V. KARTALOPOULOS, "A Bandwidth-scalable hierarchical PON for Cost-efficient Fiber Access Networks", Proceedings of the 6th WSEAS International Conference on Circuits, Systems, Electronics, Control & Signal Processing (CSECS'07), December 29–31, 2007, Cairo, Egypt, ISBN: 978-960-6766-28-2, pp. 521–526, 2007.
13. S.V. KARTALOPOULOS, "Scalable CWDM/TDM-PON network with future-proof elastic bandwidth", Proceedings of the 10th WSEAS International Conference on Communications, July 13–15, 2006, Vouliagmeni, Greece, on CD-ROM, paper 534–226, ISBN: 960-8457-47-5.
14. S.V. KARTALOPOULOS and A. SIERRA, "Engineering a Scalable and Bandwidth Elastic Next Generation PON", OFC/NFOEC 2007, March 25–29, 2007, Anaheim, CA, paper NThB4, on CD-ROM, ISBN: 1-55752-830-6.
15. ALAN HARRIS, ANDRES SIERRA, S.V. KARTALOPOULOS, and J. SLUSS, Jr., "Security Enhancements in Novel Passive Optical Networks", Proceedings of the ICC 2007 Conference, Computer and Communications Network Security Symposium, 6/24–28/07, ISBN: 1-4244-0353-7 (CD-ROM).
16. N. KOLOTHODY, A. SIERRA, and S.V. KARTALOPOULOS, "Network Protection on Next Generation PON", NOC 2005, 10th European Conference on Networks and Optical Communications, University College London, July 5–7, 2005, pp. 107–114.
17. S.V. KARTALOPOULOS, "Security and Bandwidth Elasticity Aspects of the CWDM/TDM-PON network", *WSEAS Transactions on Communications*, vol. 5, no. 8, 2006, pp. 1461–1468.
18. ALAN HARRIS, ANDRES SIERRA, S.V. KARTALOPOULOS, and J. SLUSS, Jr., "Security Enhancements in Novel Passive Optical Networks", Proceedings of IEEE ICC 2007 Conference, Computer and Communications Network Security Symposium, 6/24–28/07, on CD-ROM, ISBN: 1-4244-0353-7.
19. Z. HU, S.V. KARTALOPOULOS, and P. VERMA, "RC4-based Security in Ethernet Passive Optical Networks", Proceedings of IEEE Globecom 2006 Conference, San Francisco, on CD-ROM, NIS03-3, ISBN: 1-4244-0357-X, ISSN: 1930-529X.

20. S.V. KARTALOPOULOS, "Security and Bandwidth Elasticity Aspects of the CWDM/TDM-PON network", *WSEAS Transactions on Communications*, vol. 5, no. 8, 2006, pp. 1461–1468.

21. S.V. KARTALOPOULOS and DI JIN, "Vulnerability Assessment and Security of Scalable and Bandwidth Elastic Next Generation PONs", Proceedings of the 11th WSEAS CSCC, 7/23–28/2007, Agios Nikolaos, Crete, Greece, Advances in Communications, vol. 3, pp. 33–39.

22. FIPS PUB 140-2, *Security Requirements for Cryptographic Modules*, 2002.

23. FIPS PUB 140-2, *Security Requirements for Cryptographic Modules*, Annex A: Approved Security Functions, Draft, 2005.

24. FIPS PUB 140-2, *Security Requirements for Cryptographic Modules*, Annex B: Approved Protection Profiles, Draft, 2004.

25. FIPS PUB 140-2, *Security Requirements for Cryptographic Modules*, Annex C: Approved Random Number Generators, Draft, 2005.

26. FIPS PUB 140-2, *Security Requirements for Cryptographic Modules*, Annex D: Approved Key Establishment Techniques, Draft, 2005.

27. S.V. KARTALOPOULOS and DI JIN, "Vulnerability and Security Strategy for the Next Generation Bandwidth Elastic PONs", *Transaction of WSEAS on Communications*, vol. 6, no. 10, 2007, ISSN: 1109-2742, pp 815–823.

28. S.V. KARTALOPOULOS and DI JIN, "Vulnerability Assessment and Security of Scalable and Bandwidth Elastic Next Generation PONs", Proceedings of the 11th WSEAS CSCC, 7/23–28/2007, Agios Nikolaos, Crete, Greece, Advances in Communications, vol. 3, pp. 33–39.

29. S.V. KARTALOPOULOS, "Cascadable All-optical XOR Gates for Optical Ciphertext and Parity Calculations", SPIE Optics and Optoelectronics 2007, Prague, Chech Rep., 4/16–20/07, paper no. 6583-32, Proceedings of SPIE on CD-ROM, vol. 6581–6588.

30. S.V. KARTALOPOULOS, Next Generation Intelligent Optical Networks: From Access to Backbone, Springer, 2008.

31. S.V. KARTALOPOULOS, "Data Security in High-Speed Optical Links", Proceedings of the SPIE Defense & Security Symposium, 3/28–4/1, 2005, Orlando, FLA, session/paper number: 5814-18.

32. S.V. KARTALOPOULOS, "Terabit/s Data Communication in Distributed Computer Cluster Networks", Proceedings of the Ninth IEEE Symposium on Computers and Communications, ISCC'04 Conference, Alexandria, Egypt, June 29–July 2, 2004, pp. 1142–1147.

33. S.V. KARTALOPOULOS, "Optical Network Security: Channel Signature ID", Unclassified Proceedings of Milcom 2006, October 23–25, 2006, Washington, D.C., on CD-ROM, ISBN: 1-4244-0618-8, Library of Congress 2006931712, paper no. US-T-G-403.

34. *Optical Bit Error Rate: An Estimation Methodology*, IEEE Press/Wiley, 2004.

35. S.V. KARTALOPOULOS, "Signature analysis for channel and source authentication in optical communications", *Proceedings of the 9th WSEAS CSCC Multiconference: Communications'05*, Greece, July/14–16/2005, paper 497–226 on CD-ROM.

36. S.V. KARTALOPOULOS, "Optical Channel Signature in Secure Optical Networks", *WSEAS Transactions on Communications*, Iss. 7, vol. 4, July 2005, pp. 494–504.

37. S.V. KARTALOPOULOS, "Factors affecting the signal quality, and eye-diagram estimation method for BER and SNR in optical data transmission", Proceedings of the International Conference on Information Technology, ITCC-2004, Las Vegas, April 5–7, 2004, pp. 615–619.

38. N. SHIBATA, K. NOSU, K. IWASHITA, and Y. AZUMA, "Transmission Limitations Due to Fiber Nonlinearities in Optical FDM Systems", *IEEE JSAC*, vol. SAC-8, no. 6. August 1990, pp. 1068–1077.

39. P.S. ANDRE, L.L. PINTO, A.N. PINTO, and T. ALMEIDA, "Performance Degradations due to Crosstalk in Multiwavelength Optical Networks Using Optical Add Drop Multiplexers Based on Fibre Bragg Gratings", *Revista Do Detua*, Portugal, vol. 3, no. 2, pp. 85–90, Sept. 2000.

40. S.V. KARTALOPOULOS, "Channel signature authentication for secure optical communications", SPIE Defense & Security, 9/26–29/05, Bruges, Belgium, paper no: 5986-41, on CD-ROM: CDS191.

41. M.C. JERUCHIM, "Techniques for estimating the bit error rate in the simulation of digital systems", *IEEE JSAC*, vol. SAC-2, no. 1, pp. 153–170, 1984.

42. E. FORESTIERI, "Evaluating the error probability in lightwave systems with chromatic dispersion, arbitrary pulse shape and pre- and post-detection filtering", *Jl of Lightwave Technology*, vol. 18, no. 11, pp. 1493–1503, 2000.

43. D. MARCUSE, "Derivation of analytical expressions for the bit-error probability in lightwave systems with optical amplifiers", *Jl of Lightwave Technology*, vol. 8, no. 12, pp. 1816–1823, 1990.

44. M.D. KNOWLES and A.I. DRUKAREV, "Bit error rate estimation for channels with memory", *IEEE Trans. on Communications*, vol. 36, no. 6, pp. 767–769, 1988.

45. F. ABRAMOVICH and P. BAYVEL, "Some Statistical Remarks on the Derivation of BER in Amplified Optical Communications Systems", *IEEE Trans. On Communications*, vol. 45, no. 9, pp. 1032–1034, 1997.

46. S.V. KARTALOPOULOS, "Real-time estimation of BER & SNR in Optical Communications Channels", *Proceedings of SPIE Noise in Communication Systems*, C.N. Georgiades and L.B. White, eds., vol. 5847, 2005, SPIE, pp. 1–9. Also, invited paper at SPIE Fluctuation and Noise Symposium, May 24–26, 2005, Austin Texas.

47. S.V. KARTALOPOULOS, "Channel protection with real-time and in-service performance monitoring for next generation secure WDM networks", Proceedings of IEEE ICC 2007 Conference, Computer and Communications Network Security Symposium, 6/24–28/07, on CD-ROM, ISBN: 1-4244-0353-7.

48. S.V. KARTALOPOULOS, "In-line Estimation of Optical BER & SNR", SPIE Photon East, 10/23–26/05, Boston, Mass, Track: "Optical Transmission Systems & Equipment for WDM Networks IV", session 3, paper no. 6012-8, on CD-ROM: CDS194.

49. S.V. KARTALOPOULOS, "Per-port Circuit for Statistical Estimation of Bit Error Rate and Optical Signal to Noise Ratio in DWDM Telecommunications", Proceedings of the SPIE Fluctuation and Noise 2004 Conference, Maspalomas, Gran Canaria, Spain, May 25–28, 2004, pp. 131–141.

50. S.V. KARTALOPOULOS, *Fault Detectability of DWDM Systems*, IEEE Press/Wiley, 2001.

51. S.V. KARTALOPOULOS, "Fault Detectability: A prerequisite for Protection & Restoration in Next Generation DWDM Networks", Proceedings of the IEEE Globecom'2003 Conference, Workshop in Network Survivability, San Francisco, Dec 1–4, 2003.

52. S.V. KARTALOPOULOS, "Distinguishing between an Eavesdropper and Component Degradations in Secure Optical Systems and Networks", *Proceedings of the 9th WSEAS CSCC Multiconference: Communications'05*, Greece, July/14–16/2005, paper 497–156 on CD-ROM.

53. S.V. KARTALOPOULOS, "Differentiating Data Security and Network Security", Proceedings of ICC'08 Conference, Beijing, May 19–23, 2008.

54. S.V. KARTALOPOULOS, "Distinguishing between Network Intrusion and Component Degradations in Optical Systems and Networks", *WSEAS Transactions on Communications*, Iss. 9, vol. 4, Sept 2005, pp. 1154–1161.

55. S.V. KARTALOPOULOS, "Discriminating between faults and attacks in secure optical network", *Unclassified proceedings of Milcom'07*, 10/29–31/07, Orlando, FLA, On CD-ROM, ISBN: 1-4244-1513-6, Library of Congress: 2007904791.

56. S.V. KARTALOPOULOS, "Optical Network Security: Sensing Eavesdropper Intervention", Globecom 2006, San Francisco, on CD-ROM, NIS03-2, ISBN: 1-4244-0357-X, ISSN: 1930-529X.

57. S.V. KARTALOPOULOS and C. QIAO, "Keep your eyes open", Editorial in *IEEE Communications Magazine/Optical Communications Supplement*, vol. 14, no. 11, November 2004, pp. S2.

58. S.V. KARTALOPOULOS, "Channel Error Estimation in Next Generation Optical Networks", *WSEAS Transactions on Circuits and Systems*, vol. 3, issue 10, December 2004, pp. 2281–2284, ISSN 1109-2734, and ISBN: 960-8457-06-8.

59. S.V. KARTALOPOULOS, "Circuit for Statistical Estimation of BER and SNR in Telecommunications", Proceedings of 2006 IEEE International Symposium on Circuits and Systems (ISCAS 2006), May 21–24, 2006, Island of Kos, Greece, paper #A4L-K.2, CD-ROM, ISBN: 0-7803-9390-2, Library of Congress: 80-646530.

60. S.V. KARTALOPOULOS, "Method and Circuit for Statistical Estimation of Bit Error Rate and Signal to Noise Ratio based on Pulse Sampling for Optical Communications", USA patent no. 7,149,661, issued 12/2006.

61. S.V. KARTALOPOULOS, "Optical Network Security: Countermeasures in view of channel attacks", Unclassified Proceedings of Milcom 2006, October 23–25, 2006, Washington, D.C., on CD-ROM, ISBN: 1-4244-0618-8, Library of Congress 2006931712, paper no. US-T-G-404.

62. S.V. KARTALOPOULOS, "Next generation WDM networks with multiple channel protection and channel equalization", *IEEE 2005 Workshop on High Performance Switching and Routing (HPSR2005)*, May 12–14, 2005, Hong Kong, China, Session 4C: Resilience, Protection and Restoration II, On CD-ROM, paper no. 1568951324.

63. S.V. KARTALOPOULOS, "Secure Optical Links in the Next-Generation DWDM Optical Networks", *WSEAS Transactions on Communications*, Issue 2, vol. 3, April 2004, pp. 456–459, (ISSN 1109-2742). Also, presented at ICCC'04, WSEAS 8th International Conference on Communications and Computers, International Workshop on Cryptography, Vouliagmeni, Athens, Greece, July 12–15, 2004.

64. S.V. KARTALOPOULOS, "Optical Network Security: Countermeasures in View of Attacks", Proceedings of the SPIE European Symposium on Optics & Photonics in Security & Defense, Stockholm, Sweden, 9/11–16/2006, paper no. 6402-9; also in SPIE Digital Library at http://spiedl. org, as a part of the Optics and Photonics for Counter-Terrorism and Crime-Fighting conference proceedings: SPIE Paper Number: 6402–9.

65. ITU-T Rec. G.694.1, "Spectral Grids for WDM Applications: DWDM Frequency Grid", 2002.

66. ITU-T Rec. G.694.2, "Spectral Grids for WDM Applications: CWDM Wavelength Grid", 2002.

67. S.V. KARTALOPOULOS, "Add/Drop Capability for Ultra-High Speed Dense Wavelength Division Multiplexed Systems Using a Wavelength Bus", US patent 6,493,118, 12/10/2002.

68. S.V. KARTALOPOULOS, "Ultra-high bandwidth data transport for DWDM short, medium-haul and metro using low bit rates", APOC 2002 Conference, in SPIE Optical Switching and Optical Interconnection II, pp. 100–107, October 2002.

69. S.V. KARTALOPOULOS, "Bandwidth elasticity with dense wavelength-division multiplexing parallel wavelength bus in optical networks", *SPIE Optical Engineering*, vol. 43, no. 5, pp. 1092–1100, May 2004.

70. S.V. KARTALOPOULOS, "A Wavelength Bus Architecture for Ultra-High Speed Dense Wavelength Division Multiplexed Systems", USA patent 6,731,875, issued 5/4/2004.

71. S.V. KARTALOPOULOS, "Ultra-high bandwidth data transport for DWDM short, medium-haul and metro using low bit rates", APOC 2002 Conference, Optical Switching and Optical Interconnection II, pp. 100–107, October 2002, Shanghai, China.

72. S.V. KARTALOPOULOS, "Data Security in High-Speed Optical Links", Proceedings of the SPIE Defense & Security Symposium, 3/28–4/1, 2005, Orlando, FLA, on CD-ROM, paper 5814–18.

73. S.V. KARTALOPOULOS, "Hierarchical Encryption Technique for Dense Wavelength Division Multiplexed Systems using a Wavelength-Bus", US patent 6,577,732, 6/10/2003.

74. S.V. KARTALOPOULOS, "DNA-inspired cryptographic methods in optical communications, source authentication and data mimicking", Unclassified Proceedings of Milcom 2005, 10/17–20/05, Atlantic City, session: Comm. Security II, invited paper # 1470, on CD-ROM, ISBN: # 0-7803-9394-5.

75. S.V. KARTALOPOULOS, "A Global Multi Satellite Network", issued 2/11/1997, 5,602,838.

76. S.V. KARTALOPOULOS, "A Global Multi-Satellite Network", ICC'97, Montreal, Canada, 1997, pp. 699–698.

77. P.F. SZAJOWSKI, G. NYKOLAK, J.J. AUBORN, H.M. PRESBY, G.E. TOURGEE, E. KOREVAAR, J. SCHUSTER, and I.I. KIM, "2.4 km free-space optical communication 1550 nm transmission link operating at 2.5 Gb/s—experimental results", *Optical Wireless Communications, Proc. SPIE*, vol. 3532, 1998, pp. 29–40.

78. J.M. WALLACE and P.V. HOBBS, *"Atmospheric Science: An Introductory Survey"*, Academic Press, Orlando, 1977.

79. I.I. KIM, M. MITCHELL, and E. KOREVAAR, "Measurement of scintillation for free-space laser communication at 785 nm and 1550 nm", *Optical Wireless Communications II, Proc. SPIE*, vol. 3850, 1999, pp. 11–19.

80. S.V. KARTALOPOULOS, "DWDM: Networks, Devices and Technology", Wiley/IEEE Press, 2003.

81. S.V. KARTALOPOULOS, "Free Space Optical Mesh Networks For Broadband Inner-city Communications", NOC 2005, 10th European Conference on Networks and Optical Communications, University College London, July 5–7, 2005, pp. 344–351.

82. S.V. KARTALOPOULOS, "Free Space Optical Nodes Applicable to Simultaneous Ring & Mesh Networks", Proceedings of the SPIE European Symposium on Optics & Photonics in Security & Defense, Stockholm, Sweden, 9/11–16/2006, paper no. 6399-2; also in SPIE's Digital Library at http://spiedl.org, Advanced Free-Space Optical Communication Techniques/Applications III and Photonic Components/Architectures for Microwave Systems and Displays II conference proceedings: SPIE Paper Number: 6399-2.

83. S.V. KARTALOPOULOS, "Surviving a Disaster", *IEEE Communications Mag.*, vol. 40, no. 7, July 2002, pp. 124–126.

84. S.V. KARTALOPOULOS, "Protection Strategies and Fault Avoidance in Free Space Optical Mesh Networks", Proceedings of ICCSC'08, Shanghai, May 26–28, 2008.

85. S.V. KARTALOPOULOS, "Protection Strategies and Fault Avoidance in Free Space Optical Mesh Networks", Proceedings of ICCSC'08, Shanghai, May 26–28, 2008.

86. S.V. KARTALOPOULOS, "Security of reconfigurable FSO Mesh Networks and Application to Disaster Areas", Proceedings of the SPIE Defense and Security Conference, March 16–20, 2008, Orlando, paper no. 6975-9.

87. NIST Special Publication 800-22, *A statistical test suite for pseudorandom number generator for cryptographic applications*, May 15, 2001.

Chapter 10

Quantum Networks

10.1 INTRODUCTION

The penetration of optical fibers in communications, in concert with increased attacks on data privacy, motivated cryptographers to seek cryptographic solutions that are beyond classical theories but in the exotic, exciting and to some extend "mysterious" quantum mechanics (QM).

One of the facilitating factors in this move was the inherent quantum properties of photons, which conveniently can be exploited to create a cipher key, distribute it and create an "unbreakable" cipher text; this form of cryptography was called *quantum cryptography* (QC), and the networks that support it, *quantum networks* (QN).

An additional bonus to QC is the verification of a "mysterious" phenomenon, the *teleportation*, which if understood in depth and cost-efficiently exploited to its full potential, opens a new chapter in communications as well as in many other areas.

This chapter provides a simple description of the fundamentals of QM on which QC is based and describes the quantum properties of photons and how they have been used in QC to establish and distribute the secret key. The chapter also describes related issues with current QC methods and identifies areas for improvement. Additionally, the chapter provides a simplified description of the teleportation concept and a simplistic analog. Bear in mind that both QC and teleportation are currently topics of active research and the future is promising to bring to light many new unforeseeable developments.

The exploitation of quantum properties of photons in mathematics has also spawned another topic in research known as *quantum computation*; quantum computers promise an unprecedented computation speed (many orders of magnitude faster), although currently they do not physically exist; this topic is outside the scope of this book and chapter.

Security of Information and Communication Networks, by Stamatios V. Kartalopoulos
Copyright © 2009 Institute of Electrical and Electronics Engineers

10.2 QUANTUM MECHANICS NOT-FOR-DUMMIES

Any attempt to explain quantum mechanics in a few paragraphs or in a short chapter would be unfair to those that believe in quantum mechanics (QM) and to those that do not. However, aside from individualistic beliefs, there have been monumental theoretical developments followed by corroborated experimental evidence so that QM has shaken old intellectual and philosophical foundations, have raised eyebrows and triggered skepticism; in a manner of speaking, QM has opened a window into an unknown chapter of nature and is baffling with unexpected and difficult-to-believe phenomena.

Because some of these phenomena, when fully exploited, are expected by many to be the "holy grail" in computation and cryptography, it is educational to take a brief look into them. This chapter describes two of them, one concerning the nature of the electron and the other concerning the nature of the photon; both phenomena are described in terms of interferometry [for interesting short videoclips, please see references 1–9].

10.2.1 Electron Interferometry

Initially, electrons were thought to be particles only as a result of Niels Bohr's model of the atom, and as such when an electron gun bombards a slit on a diaphragm then electrons passing the slit form a line on a screen behind the diaphragm (Figure 10.1). Now, if two parallel slits are on the diaphragm, then instead of two lines on the screen, parallel and alternating bright and dark fringes (or an interference pattern) are seen due to constructive and destructive wave interference (Figure 10.2).

However, this result is contrary to the assumption that electrons are small particles only, and it can only be explained if it is assumed that electrons are also waves, since the experiment with waves and two slits had already been confirmed. Bright

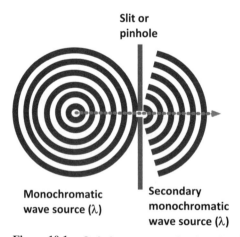

Slit or pinhole

Monochromatic wave source (λ)

Secondary monochromatic wave source (λ)

Figure 10.1. Optical wave propagating through a slit (left to right).

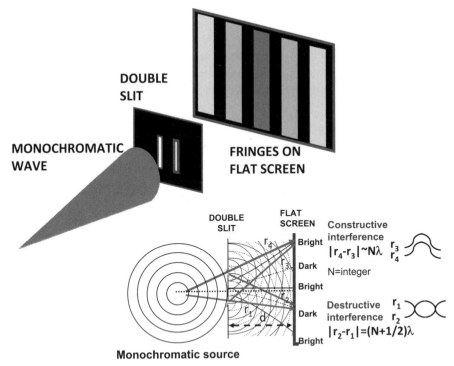

Figure 10.2. Optical wave propagating through a double slit forms an interference pattern on a screen behind it.

and dark fringes are explained as the result of superimposing two waves of the same frequency; when they are in phase, the waves are added producing bright fringes (known as constructive interference) and when the two waves are in opposite phase (that is, displaced by 180 degrees), the waves cancel each other, producing dark fringes (known as destructive interference).

The skeptics of this experiment thought that when electrons bombard the two slits, there must be an interaction between them and it is the interacting forces that make them behave like waves before they hit the two slits, hence the fringes on the screen. So, the experiment is repeated, but this time with a single electron gun, one electron at a time to eliminate possible electron wave interactions, and the astonishing result from this experiment was that the fringes did not disappear!

The conclusion from this experiment could now only be explained if one accepts the assumption that electrons must behave like waves as well, known as the *duality nature* of electrons (for an eloquent and humorous animation on this topic see http://www.youtube.com/watch?v=DfPeprQ7oGc).

The duality nature was also a convenient explanation because now the Bohr's model that considered electrons to be particles orbiting the nucleus of an atom at any distance, could be reformulated and explain some previously unexplained observations. This simple model was expanded to a more complex model that supported, among other ideas, that:

- electrons are described by density wave functions (Erwin Schrödinger),

- an electron's position and momentum in space cannot be simultaneously and precisely measured, yet the product of uncertainty of both is fixed $\Delta x \Delta p = h/2\pi$ (Heisenberg),

- electrons are at specified orbits from the nucleus and not at any distance from it; that is, the orbiting electrons are at a quantized distance from the nucleus (see Bohr's formula below), and

- between nucleus and orbiting electrons are forces and *nothing else* (nothing else means *emptiness* which is difficult to define).

In all, there is a theory declaring that subatomic scales are quantized and that the world is not 3-dimensional linear according but 4-dimensional curvilinear, where the fourth dimension is time (Einstein); Einstein in his first paper solved the first major problem in physics, the light quantum, and quantum theory has been a successful scientific theory based on its consistency with a broad range of experimental observations.

In fact, Heisenberg's paper "Quantum-mechanical re-interpretation of kinematic and mechanical relations" in 1925 launched quantum mechanics. He used matrix multiplication, in which case $A \times B \neq B \times A$, establishing the *non-commutativity* of kinematical quantities; that is, measuring first the position and then velocity would not produce the same result as measuring first the velocity and then position.

The experiment with the single electron gun was repeated, but this time one detector was placed near one slit to count the number of electrons that passed through in an attempt to understand the problem statistically. The experimental results behind the two slits went back to two crisp lines without showing any fringes.

This was an unexpected result that led scientists to another conclusion: the mere measuring of electrons before the slit must have influenced the elecrtron's wave nature and made them behave like particles only, as if the electrons knew that someone was watching them.

10.2.2 Photon Interferometry

The initial model of light considered light to be a pure electromagnetic wave that propagated in free space at the "speed of light," which is approximately $c = 10^{10}$ *cm/sec*. As pure wave, light verified all known experiments of wave mechanics at the time, including interferometry.

Based on Bohr's model, the classical view of the photon postulated that photon's energy is proportional to its frequency, $E = h\nu$, where ν is the frequency and h is Plank's constant. When an electron from a high energy level E_1, which is associated with a higher energy orbit, transits to a lower energy level E_2, then the difference of energy is emitted as energy $\Delta E = E_1 - E_2 = h\nu$, which can be photonic. In addition, photons are mass less, charge less, spin less and they exert pressure according to their momentum relationship, $E/c = p$.

Later, Bohr generalized Balmer's formula for hydrogen and described the wavelengths λ that are generated between orbits as:

$$1/\lambda = R_H(1/m^2 - 1/n^2),$$

where $R_H = 109,677.58\,cm^{-1}$ is the Rydberg constant for the Hydrogen atom and n and m are integers $n,m = 1,2,3,4,5 \ldots$ with $n > m$. That is, this relationship is quantized since n and m are integer numbers and not continous, and it also also explains the spectral lines of atoms that the simple model cannot.

To shorten the story, in subsequent years, the assumption of the pure wave as the nature of photons also changed and the *dual nature* was accepted, wave and particle. The question now was: do photons behave like the electrons do at an interferometer? Do they know that they are watched? The problem with this is that photons do not have charge and thus a charge detector cannot be used to sense passing by electrons. If a photodetector was used, it would absorb photons and the experiment would not produce good results. To address this, a different mechanism was used, known as Mach-Zehnder interferometer.

According to it, a photon source emits a beam that consists of many coherent photons (that is, photons that are in phase). This beam is split in two beams, A and B, by a 50-50 splitter, S_1, so that each beam travels a separate path (path A and path B). Each of the two beams is reflected by a mirror, M_A and M_B, respectively, and the two beams A and B arrive at a second beam splitter S_2, which splits each one 50-50, and behind this beam splitter there are two detectors D_1 and D_2 (Figure 10.3). The assumptions in the experiment are that the two path lengths are identical so that there is no phase difference introduced due to unequal travel length. In addition, one needs to consider that each time a wave is reflected by a mirror, its phase shifts by 180 degrees.

So the question now is: based on interferometry, what can be predicted for the two detectors D_1 and D_2?

Counting the total phase shift from source to S_2 over path A, the net phase shift is 180 degrees and thus D_1 should not detect anything due to destructive interference, and only detector D_2 should detect light as a result of constructive interference, a result that is verified experimentally.

Now, what if the experiment is repeated with a single photon source, like the single electron gun experiment, one photon at a time?

In the experiment with single photons, one would expect that the photon follows either path A or path B with 50% probability, and when the photon arrives at splitter S_2 then the photon is either detected by detector D_1 or by D_2, again with 50% probability since there are no two photons to interfere. What has been reported as experimental evidence, however, is not this; the result reported is that only D_2 detects photonic activity!

So, what does this unexpected result prove?

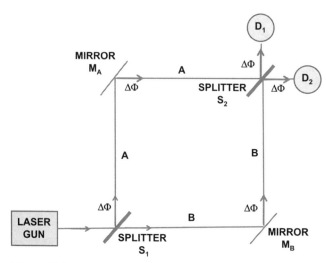

Figure 10.3.   The Mach-Zehnder interferometer.

To understand the significance of this experiment and answer this question, one has to look into the *superposition principle* rooted in quantum mechanics, on which quantum computation and quantum cryptography is based.

The superposition principle starts from the premise that, in quantum mechanics, a particle that can have specific values, such as electron spin that can be $+1/2$ or $-1/2$, has a quantum state that is the superposition of the two until the state of spin is measured. This superposition of states is expressed in terms of density functions as the sum of probabilities of the two states. Because the probability to be in one state is equal with the probability to be in the other, the two probabilities are the same (0.5 each, so the sum is 1).

In classical mechanics there is no analog to the superposition principle because classical mechanics provides a macroscopic view and not a sub-atomic view as QM does; in CM, the spin value is either $+1/2$ or $-1/2$ and nothing more.

Based on the superposition principle, the following obvious explanations can be given: the photon wave nature is split in two waves, one traveling along path A and the other along path B. When the two waves recombine at S_2, then due to constructive interference at S_2, photonic energy is detected by D_2 and not by D_1. However, there are few small problems with this:

A. If the wave is split, the energy is also split. This would have produced two photons with half the frequency as a result of the energy conservation principle.

B. What happened to the particle nature of the photon? Is this also split along with the wave nature? This nature does not have mass and thus the conservation of mass is not violated. Was this nature converted to pure wave?

C. What do photodetectors detect? The wave nature or the particle nature of photons? Clearly they detect photonic energy as a result of the photoelectric effect (which also has roots in QM).

These are good questions, but a plausible explanation to this result is the super-position principle of QM. The photon has two choices when at splitter S_1. Representing two states, and although the photon may follow path A or path B, the superimposed states travel along with it. However, when at splitter S_2, the superimposition collapses and the states recombine yielding constructive interference at D_2.

10.2.3 Convergence of Two Radical Ideas in Quantum Mechanics

Heisenberg and Schroedinger envisioned the radical ideas that a subatomic particle (electron) is not like a well-defined tennis ball but a density at the center of a disturbance and they converged to the same quantum mechanical definition of a single electron from two different approaches.

- Schroedinger used a wavefunction $<\psi>$ and probability distributions to describe the wave nature of a single electron. He considered that this density spreads in space like a wavy cloud and thus the electron is somewhere in it with some probability. He described the pure state.

- Heisenberg described the uncertainty of locating a single electron particle. He considered the particle-like electron having a density with probabilistic properties that are formulated with matrix manipulation.

The convergence of the two theories was made by Max Born who discovered that the square of Schroedinger's wavefunction, $<\Psi>^2$, computes the electron's probability distribution. That is, Born located the electron, starting from Schroedinger's wave function, somewhere in space with probability of being here more than there according to Heisenberg's probability distribution function.

Hence, the remarkable discovery that the electron is described with density wave functions, with probabilities, and with uncertainties. When the exact position of an electron is measured with a device not in terms of a probability distribution but in terms of its physical position, the electron's wave-like properties sharply diminish due to the electron-detector interaction, the electron behaves like a well-defined particle, and Schroedinger's equation fails. This is known as the *wavefunction collapse*.

10.2.4 Superposition of States, Qubits and Qu-Registers

A particle's state in quantum mechanics, based on Schroedinger, is a state vector. This state vector is denoted as $|\psi>$; the new symbol "$|>$" is called *ket*, and the symbol "$<|$" *bra*; hence, $<|>$ is a *bra-ket*; this bra-ket notation has been coined by Dirac and it has been used ever since.

Based on Dirac's notation, the Schroedinger's equation is written in the time partial differential form of state vector |ψ>:

$$i(h/2\pi)\partial|\psi>/\partial t = \boldsymbol{H}|\psi>,$$

where h = Plank's constant, and $\boldsymbol{H}$ is the Hamiltonian operator.

If $\boldsymbol{H}$ is not time-dependent, then the wavefunction solution of Schroedinger's equation (for which Heisenberg's uncertainty $\Delta x\Delta p > h/2\pi$ still holds) is:

$$|\psi(t)> = e^{-iHt}|\psi(0)>$$

Now, the probability distribution p_m of a measurement of a standard observable quantity A performed in state |ψ> (and based on Born's discovery) is the square of Schroedinger's wavefunction:

$$p_m = |<a_m|\psi>|^2.$$

Similarly, the probability distribution p_m of the measurement results of a generalized observable quantity $\{\boldsymbol{M}_m\}$ that is performed in state |ψ> is:

$$p_m = |<\psi|\boldsymbol{M}_m|\psi>|^2.$$

In classical mechanics, the binary state of a bit has the undisputable value "0" or "1." In QM, if the state of a wave particle has not been measured, then the wave particle is in one state with probability p and in the other state with probability $1 - p$; the exact state will be known only when the measurement is made.

Notice that results of quantum measurements do not state pre-existing properties (there is no memory). In addition, the state vector |ψ> simultaneously determines the probability amplitudes at all locations; that is, |ψ> → ψ(r), for all r (the word *simultaneously* here hints superposition). That is, in a coordinate space representation, the Schroedinger equation is a linear partial differential with a special property: the linear superposition of a set of unique solutions of the equation (also known as eigenfunctions) is also a solution of the equation. The notion that both states of a binary system exist simultaneously is the outcome of Schroedinger's equation and it is known as the *superposition state*; for a binary system, the classical mechanical bit becomes in quantum mechanics a "*qubit*".

Thus, assuming a balanced system, then $p = 1/2$; that is, the probability of being in one state or in the other is 50%. This may be paralleled with tossing a coin: as the coin spins fast in mid-air, one does not know in which state it is, head or tail; thus, it is considered to be in a superposition of states with state probability |ψ> = (1/√2)(α|1 > +β|0>), where $\alpha^2 = \beta^2 = 0.5$ (that is, each state head or tail has equal probability). In this simplistic experiment, when the coin lands, its actual state is measured and is 100% known, and yet there is 50% probability of landing in one state or the other.

Qubits may represent the spin states of an electron (these are two eigenstates) or two polarization states of a photon. If the *spin +½* is associated with logic "1," and the *spin –½* with logic "0," using Schroedinger's notation this is written as:

$$|1> = |+ \tfrac{1}{2}>, \text{ and}$$
$$|0> = |- \tfrac{1}{2}>.$$

Similarly, the photon polarization ↑ or ↓ associated with logic "1" and logic "0," respectively, is denoted as:

$$|1> = |↑>, \text{ and}$$
$$|0> = |↓>.$$

Following the same (ket) notation, the superposition property can now be written as a superposition of eigenstates of the *qubit*, which (for a binary system) is:

$$|\psi> = \alpha|1> + \beta|0>.$$

where α and β are constants related with the probability $(|\alpha|^2)$ of being in state "1" and with the probability $(|\beta|^2)$ of been in state "0", such that $|\alpha|^2 + |\beta|^2 = 1$.

Having defined the qubit and its notation, a quantum n-bit register (qu-register) is written as:

$$|\psi> = |1>_1 \otimes |1>_2 \otimes \ldots \otimes |1>_n$$

In addition, a unitary operator is defined such that $U*U = 1$ (this becomes useful when normalization is needed), and also other operators, such as for example the U_{NOT} (its classical analog is the gate NOT) as well as other operators that have no classical analog, which that are useful in *quantum logic* and *quantum computation* (which is beyond our objective).

If the U_{NOT} operator corresponds to flipping spin states from $+\tfrac{1}{2}$ to $-\tfrac{1}{2}$, then it operates on the two states as follows:

$$U_{NOT}|1> = |0>, \text{ and}$$
$$U_{NOT}|0> = -|1>.$$

Furthermore, this can be generalized as a transformation of superposition of states to a different superposition, as:

$$U_{NOT}\left\{ (1/\sqrt{2})(|1> + |0>) \right\} = (1/\sqrt{2})(|0> - |1>).$$

A transform that takes a basis state and transforms it into a linear combination of two basis states is known as the "Hadamard" or as the "Walsh-Hadamard" transform; the Hadamard transform is a special case of the more general Fourier transform and it is useful in quantum algorithms, as follows:

The general Hadamard transform H_m is a $2^m \times 2^m$ matrix that transforms 2^m real numbers in x_n to 2^m real numbers in X_k. A 2×2 general Hadamard matrix is:

$$H_m = (1/\sqrt{2}) \begin{vmatrix} H_{m-1} & H_{m-1} \\ H_{m-1} & -H_{m-1} \end{vmatrix}$$

where $1/\sqrt{2}$ is normalization constant. For $m = 1$, all matrix elements are $H_0 = 1$ and the matrix is made up of elements $+1$ and -1.

In a more general case, the kth and nth entry in the Hadamard matrix are:

$$k = k_{m-1}2^{m-1} + \ldots + k_1 2^1 + k_0, \text{ and}$$

$$n = n_{m-1}2^{m-1} + \ldots + n_1 2^1 + n_0$$

where the terms k_j and n_j are binary digits 0 or 1.

In this case, $H_m|k,n = (2^{m/2})(-1)\Sigma_j k_j n_j$, which is the multi-dimensional $2 \times 2 \times \ldots \times 2$ discrete Fourier transform (DFT).

Now, the Hadamard transform (with $m = 1$) is

$$H_1 = \left(1/\sqrt{2}\right) \begin{vmatrix} 1 & 1 \\ 1 & -1 \end{vmatrix}.$$

This, in quantum theory is the rotation transformation of qubit-basis states $|0>$ and $|1>$ to two superposition states with equal weight of the computational basis states $|0>$ and $|1>$, which using Dirac's notation is written:

$$\left\{(|0> + |1>)/\sqrt{2}\right\} < 0| + \left\{(|0> - |1>)/\sqrt{2}\right\} < 1|.$$

One of the transforms (but with no classical analog like the NOT) is the "square NOT", U_{SRN}, defined by Deutsch [10–13]:

$$(U_{SRN})^2 = U_{NOT}$$

The U_{SRN} performs transformations on quantum states, which in physical terms correspond to spin rotation by 90 degrees:

$$U_{SRN} |1> = \left(1/\sqrt{2}\right)(|1> + |0>)$$
$$U_{SRN}|0> = \left(1/\sqrt{2}\right)(-|1> + |0>).$$

This transformation on a qu-register generates a superposition of all 2^n states:

$$|\psi_n>' = U_{SRN} \otimes U_{SRN} \otimes U_{SRN}|11\ldots 1>$$
$$= \{1/(2^{n/2})\}(|1> + |0>) \otimes (|1> + |0>) \otimes \ldots \otimes (|1> + |0>)$$
$$= \{1/(2^{n/2})\}\{|11\ldots 1> + |11\ldots 0> + \ldots + |11\ldots 0>\}.$$

The final conclusion is that by applying a linear number of operations to the n-bit qu-register, a register state with 2^n terms is generated. It is this superposition of states that gives the power to quantum parallel computing and to cryptography. Add to this the quantum entanglement, and then something very powerful in terms of computing power takes place that has led many to believe that quantum computation is incredibly fast and also that quantum cryptography is unbreakable.

10.3 QUANTUM CRYPTOGRAPHY

Before entering this subject, it is necessary to clarify that the two-word term "quantum cryptography" does not mean that cryptography is quantized or that quantized quantities are cryptographic. However, it describes a technology that uses quantized properties and exploits quantum mechanical principles and phenomena such as superposition of states, interferometry, and entanglement and a sophisticated algorithm to establish and distribute a secret code from point A (source or Alice) to point B (the receiver or Bob) such that neither the B knows the key nor anyone else in a middle point E (Evan).

The communications channel that is most suitable to quantum cryptography is the optical; the optical fiber medium is most widely used, although there are applications in free space optical links. The reason for optical links is that photons can maintain their quantum properties in communications, whereas free electrons in copper wires do not maintain their spin, and hence the quanto-mechanical advantage in optical communications.

Experiments in quantum cryptography exploited the Mach-Zehnder interferometer, similar to that in Figure 10.3, with the difference that on the path A and B there are two controllable phase shifters. Controlling the phase shift on each path, detector D_1 or detector D_2 is able to detect constructively interfering photons.

The following section focuses on fiber optic communications and on the most popular photon property, polarization (qubits and entangled states). For a description of several methods with quantum-mechanical notation see reference [14].

10.3.1   Pre-Quantum Cryptographic Era

The cryptographic systems described so far belong to classical cryptography. Such systems consist of a source A (Alice), a communications channel or link, and a receiver B (Bob) (Figure 10.4).

- The source includes a cryptographic module that contains or derives the encryption keys, encrypts the plain text and it transmits the cipher text.

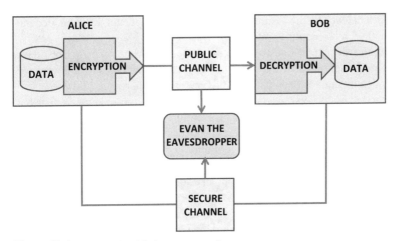

Figure 10.4.   Classical public key cryptography.

- The receiver receives the cipher text and passes it through a de-cryptographic module that contains or derives the decryption keys outputting the plain text.

- The channel or link is the medium which may be considered to be tapped by an eavesdropper, Evan.

According to public key cryptography, without prior agreement, Alice and Bob randomly create their encryption/decryption key-pairs, these key-pairs are mutually inverse transformations. Alice then, using a conventional public communication channel, publishes the instructions for encryption but not decryption.

Bob uses Alice's public-encryption algorithm to prepare a cipher text that only Alice can decrypt. Similarly, Alice and anyone else can use Bob's public encryption algorithm to prepare a message that only Bob can decrypt.

Thus, Alice and Bob can exchange secret messages without sharing the same key.

As already described, classical cryptography security relies on the difficulty of breaking the key, which is rooted in large prime numbers as already described.

10.3.2   Post-Classical Cryptographic Era

Quantum cryptography (QC) uses the power of quantum mechanics that promises the distribution of a secret key to remain secure even if Evan has an unlimited computation power. This may sound like a paradox, because if Evan has an unlimited computational power at his disposal, then nothing can stand in his way! However, this text has described many phenomena that Einstein and others used to call "weird," and these are the ones that QC exploits.

For example, quantum bits is the coherent superposition of quantum states and any attempt to measure, copy or clone the states will destroy the superposition state, and it will measure one of the states at random; that is, the outcome of a single measurement is unpredictable. Hence, whereas a classical logic bit remains unchanged by whatever method is inferred, a quantum bit appears different, depending on the method by which it is inferred.

Now, assume an optical link between two points A (Alice) and B (Bob). How can Alice create a quantum key that only she knows and distributes to Bob who will not know its details but who can use it to decipher the ciphertext?

The method that answers this question, generating a quantum key and distributing it to Bob, is known as "quantum key distribution" (QKD).

Many quantum cryptographic methods, and particularly QKD, use polarized photons. According to it, a state or a subset of photon polarization states corresponds to logic "0," whereas another state or a subset of states corresponds to logic "1;" this becomes evident if one consults the polarization states on a Poincaré sphere, Figure 10.11.

The polarization states and their logic association are initially known to Alice. Through a process explained below, Alice defines the encryption or quantum key and makes it (partially) known to Bob. Thus, the secrecy of this method and the encryption algorithm promises a secure communications channel.

10.3.3 Photon Polarization States

The polarization of light is a well-known phenomenon, particularly to those in photography and to those who wear sunglasses. This simplistic statement may sound very vague and un-insightful to quantum mechanics enthusiasts. However, this is the classical view of polarization of light and is still in use.

According to the classical wave theory, light is an electromagnetic wave such that its electric and magnetic fields are perpendicular to each other. When an excited atom (atom at a higher energy) drops to a lower energy, under certain circumstances, the difference of energy is converted to photonic energy, $\varepsilon = \varepsilon_1 - \varepsilon_2 = h\nu$, and a photon is born with a magnetic field that remains orthogonal to electric; Figure 10.5 is an attempt to capture the moment a photon is born given the fact that electrons travel on orbits around the nucleus on which their exact location is not precisely known. Hence, they travel on fuzzy trajectories.

However, to a stationary observer the two fields evolve around each other and both move in a direction that is perpendicular to the plane the two fields define (Figure 10.6). The sinusoidal wave motion of light is mathematically described by Maxwell's equations.

When light passes through a polarizing film, the charge distribution in the film interacts with the electric and/or magnetic field of light more in selected angles than others. The end result is that the exiting light has a field that is stronger in the orientation of the polarization axis of the film, as this is the least orientation that

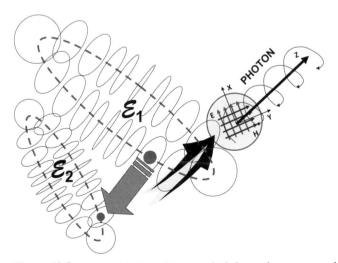

Figure 10.5.   A photon is born when an excited electron loses energy and transits to a lower energy level.

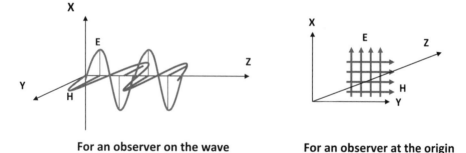

Figure 10.6.   Classical view of electro-magnetic wave propagation.

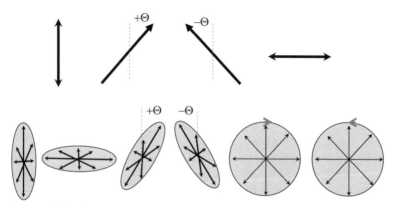

Figure 10.7.   Classical view of a linear, elliptic and circular polarization.

influences the field. In general the output of a polarizer is linear, elliptic, or circular (Figure 10.7).

As a consequence, if a linearly polarized wave passes through another linear polarizer with perpendicular axis of polarization to that of light, the result is devastating because light will not get through the film. If the axis is not perpendicular but in an angle θ, then the amount of light getting through the film is $\sin^2\theta$ times the amount of light that entered the film (Figure 10.8).

To avoid confusion in subsequent sections it is important to explain that what really takes place is that a beam of polarized light consists of many same polarized photons that arrive at the film. Then, probabilistically, $\sin^2\theta$ of photons will pass through and $\cos^2\theta$ will be absorbed. Now, what if there is no beam but only a single photon? Similarly, the probability that the photon will pass through is $\sin^2\theta$ and the probability that will not pass through is $\cos^2\theta$. That is, there is no way of knowing if the photon will pass or will not pass but what the probability is that it will or will not.

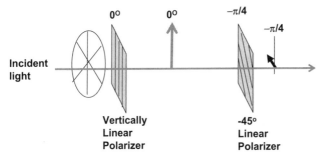

Figure 10.8. Polarized light at an angle with the polarization axis of a polarizer.

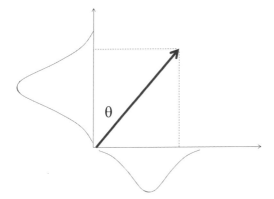

Figure 10.9. Probability distribution of a general polarization when at an angle with a polarizer.

10.3.3.1 Orthogonal States

Interestingly enough, this chapter started from a classical viewpoint and ended up with a probabilistic one, which is clearly much closer to quantum mechanics. So what really happens?

With respect to the polarizing film and the probabilities, the quantum mechanical viewpoint is that a polarized photon in an arbitrary angle is a vector that may be viewed as consisting of two states coexisting, a ↑ state and a → with probabilities α and β for each state, respectively, so that their sum is 1. Thus, when the photon arrives at the polarizer, depending on the polarization orientation of the polarizer axis, there is a probability that one state will pass and another probability that the other state will, but it is unknown which of the two actually passes. If the axis coincides with one of the two polarization components of the photon, then that component has probability 100% (Figure 10.9).

The general quantum mechanical view of photon polarization is that when a photon is generated it has a left-hand circular polarization state (L) or a right-hand circular (R) with equal probability [15]. That is, both states coexist and the actual state is not known until it is measured. Each state is associated with the classical logic "1" and the classical logic "0," and this sets the starting point of quantum cryptography.

A slightly different quantum mechanical view is that single photons, like electrons, are wave-particles that possess an intrinsic angular momentum; however, the electron possesses resting mass whereas the photon is a relativistic wave-particle. Thus, for the electron the intrinsic angular momentum is spin with states $+\frac{1}{2}$ and $-\frac{1}{2}$, whereas for the photon the intrinsic angular momentum is called helicity or circular polarization [16]. According to this, the polarization is associated with a polarization angular momentum quantum number $j = 1$ with the two polarization base states given by $|m_j = +1>$ for the right-hand circularly polarized state and $|m_j = -1>$ for the left-hand circularly polarized state, and because of photon's relativistic nature, the state $m_j = 0$ does not exist. That is, the photon's state of intrinsic angular momentum parallel to its direction of motion can only be $+1$ or -1 (in units of $h/2\pi$), and there is no component of intrinsic angular momentum perpendicular to the direction of motion [17]. Thus, even this view also converges to the quantum mechanical right-hand/left-hand polarization state of created photons.

Using quantum mechanics Dirac notation of a pair of ortho-normal states, such as vertical polarization V or $\updownarrow$ and horizontal H or $\leftrightarrow$, then the right-hand (R) and left-hand (L) polarizations are written:

$$|R> = (1/2\pi)(|H> + i|V>), \text{ and}$$

$$|L> = (1/2\pi)(|H> - i|V>).$$

Similarly, one may think as two pair of two diagonal and ortho-normal states, such as 45° or / and 135° or \, and so on.

Because the two states $|V>$ and $|H>$ are orthogonal to each other (orthonormal), then it follows that $<V|V> = <H|H> = 1$ and $<V|H> = <H|V> = 0$.

In addition, if the photon is in a known polarization quantum state $|\psi>$ and it ends up in a polarization quantum state $|\chi>$ as a result of something done to it (such as an interaction or rotation of polarization), then $<\chi|\psi>$ is the quantum probability amplitude that the photon will go from state $|\psi>$ to $|\chi>$. Then, the probability measure that the photon begins in $|\psi>$ and ends up in $|\chi>$ is given by the absolute square of the quantum probability amplitude:

$$\textbf{\textit{Probability}} = <\chi|\psi><\chi|\psi>^* = <\chi|\psi><\psi|\chi> = |<\chi|\psi>|^2$$

10.3.3.2 *Creating Orthogonal States*

The exploitation of quantum mechanics in cryptography requires the creation of two orthogonal states. In photonics, this is not a difficult task if a strong bi-refringent crystal is used, such as calcite.

It is known that when light enters such a crystal in a specific direction (with respect to the crystallographic axis), the light beam is split into beams following two directions along which the dielectric constant is different and thus the index of refraction and the propagation properties of the two beams [18]. As it turns out, each

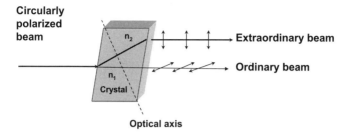

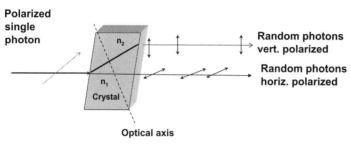

Figure 10.10.   Calcite is a bi-refrigent crystal and has been used in quantum cryptography experiments. If the crystal is properly oriented, horizontal polarization passes through unchanged (ordinary), vertical polarization is spatially shifted and the polarization remains (extraordinary), and polarizations at 45° are passed through with horizontal (ordinary) or vertical (extraordinary) polarization with equal probability.

beam is polarized differently in one crystallographic axis than the other axis. One beam is known as ordinary and the other as extraordinary, and the two states are known as entangled states [19, 20]

Now, if a single photon enters such a crystal, the crystal does not create two photons (in photon entangled states we will also discuss under what circumstances this can be done) but it creates one of two polarization states with equal probability of being in one or the other; however, as long as the photon is in the crystal, it is a qubit because the actual state is not known, similar to a coin in mid-air.

When a photon passes through a calcite crystal, the photon may emerge polarized perpendicularly to the crystal's optic axis, or it may be shifted and emerge polarized at 45° or 135°. More specifically, the calcite crystal works as shown in Figure 10.10.

10.3.4   Photon Polarization States

10.3.4.1   The Poincaré Sphere

At this point, it is important to take a step back to describe in classical and practical terms the mapping of polarization states of photons on the surface of a sphere (in spherical coordinates (r, θ, ϕ), known as the *Poincaré sphere*. The Poincaré sphere provides a visualization of photon polarization states, where the states $\uparrow$, $\rightarrow$, $\downarrow$, $\leftarrow$,

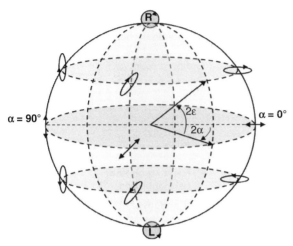

Figure 10.11.   The Poincaré sphere helps to visualize the polarization states of photons.

|*R*> or |*L*> are individual states mapped on surface points whereas all other points on the sphere map other diagonal linear and elliptical polarization states (Figure 10.11).

A polarization shifter is a device that changes one state of polarization to another. For example, the polarization state can change from ↑ to →, or from |*R*> to |*L*>, and so on. As a state of polarization (SoP) changes from one to another, and if the change is continuous, then a trace is scribed on the surface of the Poincaré sphere from the starting SoP to the ending SoP, unless the change is abrupt, in which case there are only two points, the before and after SoP.

10.3.4.2 Rotating States of Polarization

A state of polarization may be rotated with one of several methods. One method uses a rotating polarizer, and another method uses an electro-optic device based on the Pockel effect. Because the Pockel device rotates the polarization state extremely fast by applying a voltage and does not require mechanical means, it is more popular.

The Pockel effect considers the refractive index (which is related to the dielectric of material) as a series terms expanded as the applied electric field $n = n_0 + a_1E^1 + \ldots$ However, it considers a class of materials for which the coefficient α_1 is significantly large and higher order terms insignificantly small and thus $\Delta n = a_1E$; such materials are certain crystals such as GaAs, LiNbO$_3$ and KDP (KH$_2$PO$_4$-potassium dihydrogen phospate).

10.3.4.3 Qubit Basis States

Two orthonormal polarization states such as the rectilinear states that consist of a vertical 0° polarization and a horizontal 90° (↕ and ↔) polarization is known as basis. The diagonal of 45° and 135° (/ and \) is another basis, and so on.

In general, a qubit can be represented as a point on the surface of a sphere defined by state vectors in the S_x, S_y, and S_z, the radius of which is:

$$R_0 = \sqrt{\left(S_x^2, S_y^2 + S_z^2\right)} = 1$$

This sphere is known as the Bloch sphere which in a spherical coordinate system (r, θ, ϕ) is the Poincaré sphere. Transforming a qubit to another qubit by performing three rotation transformations about the three axes in the Cartesian coordinate system (x, y, z) will move the original point on the sphere to another.

In QC, according to the selected bases, polarization states are associated with a binary bit (0 and 1) based on a schedule such as 0 corresponds to the state $\updownarrow$ or /, and 1 corresponds to $\rightarrow$ or \. That is, one may select the qubit basis ($|0>$ and $|1>$), although in theory any other basis may be selected. In general, any unitary transformation of the basis vector will produce a valid new basis set. For example, the original set $|0>$ and $|1>$ represented by a point $\pm e_z$ on the Poincaré sphere, after the Hadamard matrix $H = (X + Y)/\sqrt{2}$ transformation generates the new basis set:

$$|\varepsilon_0> = \left(1/\sqrt{2}\right)(|0> + |1>), \text{ and}$$
$$|\varepsilon_1> = \left(1/\sqrt{2}\right)(|0> - |1>).$$

However, one should remember that the eigenstate $|1>$ or $|0>$ in basis ($|0>$ and $|1>$), after the transformation, is no longer an eigenstate in the new basis ($|\varepsilon_0>$ and $|\varepsilon_1>$).

10.3.4.4 *Rules of Quantum Cryptography Algorithms*

The initial idea to use quantum physics for data privacy was Stephen Weisner's in the early 1970s, when he authored at a paper titled *Conjugate Coding* while at Columbia University. In it, he supported that quantum physics can accomplish a way to produce bank notes that would be physically impossible to counterfeit, and also combine two classical messages into a single quantum transmission from which the receiver could extract either message but not both. This paper was not accepted by the journal to which he had submitted it, and it remain unpublished until 1983.

However, his paper was later researched and eventually it was published in the *Sigact News* by Charles Bennett and Giles Brassard of IBM Thomas J. Watson Research Laboratory who also built upon Weisner's ideas.

Based on superposition of polarization states of photons or qubits, circular $\{|L>, |R>\}$ or linear $\{|\uparrow>, |\rightarrow>\}$, such that $|R>$ represents logic "0" and $|L>$ logic "1," Bennett, Breidbard and Brassard realized that a method could be devised so that a secret key can be established between A and B, which can be known only to Alice. Bob can use it even if he does not know the key and Evan cannot intercept it.

Thus, in 1982 they introduced the first method for quantum cryptography using qubits [21], and subsequently in 1984 they developed the first quantum cryptography protocol, known as BB84 [22, 23], and in 1991 a proof-of-concept prototype.

Since then, quantum cryptography has followed a rapid development so that to date there are several protocols and at least two major companies that provide system solutions in the U.S., in Europe and elsewhere, whereas others are scrutinizing the security aspects of it [24–34].

Besides individual photons and qubit states, quantum cryptography depends on Heisenberg's uncertainty principle, according to which measuring a quantum system disturbs it and yields incomplete information about its state before the measurement. In sort, quantum cryptography is based on the indeterminacy of qubits, as follows:

- The state of a qubit cannot be detected or measured without disturbing it.
- A qubit cannot be copied without disturbing it.
- A qubit cannot be cloned without disturbing it [35].
- Any attempt (by an eavesdropper) to measure, copy or clone a qubit, destroys its superposition state and this becomes noticeable at the receiver.

10.3.5 The BB84 Protocol

The BB84 protocol capitalizes on the aforementioned ideas and on photon polarization states to transmit data in privacy between Alice and Bob via an optical quantum communication channel. The transmission medium can be free space or silica-based optical fiber. In addition to the quantum channel, communication also takes place via a public channel, wired or wireless of any public network (PSDN or the Internet). Neither of these two channels need to be secure; the assumption is that an eavesdropper, Eve or Evan, has gained access to both channels (Figure 10.12).

BB84 uses two conjugate bases, the rectilinear basis of vertical 0° and horizontal 90° (↑ and →), and the diagonal basis of 45° and 135° (/ and \); a basis is a pair of

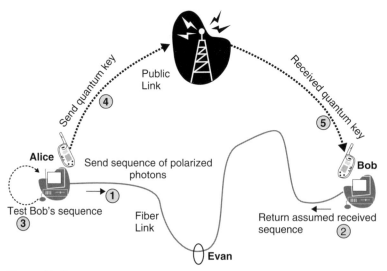

Figure 10.12. Quantum networks use a quantum channel (optical) and a public (wired, wireless) channel.

orthogonal states. Polarization states are associated with a binary bit (0 and 1) based on a schedule such as 0 corresponds to the state ↑ or /, and 1 corresponds to → or \.

Based on this, the secret key between Alice and Bob is established as follows:

- Alice uses a polarization filter right after the photon source, which transforms her binary data string of "1s" and "0s" symbols in a string of polarization states by randomly selecting one of two states for each logic symbol. For example, logic "1" may be represented by → (0°) or by \ (135°) that are randomly selected, and logic "0" by / (45°) or ↑ (90°), also randomly selected; the bases and polarization sequence is known to Alice only and no one else, including Bob.

- As each photon in the data string is polarized, it is coupled onto the fiber and transmitted over the quantum channel to Bob.

- Bob does not yet know the bases and randomness that Alice had selected, so he selects his own basis at random (rectilinear or diagonal polarization). Based on his selection, he passes each arrived photon through his polarizer behind which is a photodetector.

- Bob records the bit sequence and the measurement basis he used and also the result of the photodetector (a polarized photon that coincides with the polarization axis of Bob's polarizer will pass the polarizing filter and the photodetector will produce an output, else the photon will be blocked and the photodetector will produce no output).

- As soon as Bob has measured the last photon, he communicates with Alice over the public channel and tells her which basis, rectilinear or diagonal, he used for each photon he received but he does not tell the results of his measurements.

- Alice performs a test using Bob's basis and tells him which photons are common between his measured ones and her transmitted ones. Alice may also discuss with Bob the bases she used.

- Alice and Bob keep the data that pass correctly and discard the rest that correspond to Bob's wrong measurements.

- The correct data in this protocol establish the key, if no one has eavesdropped the quantum channel (Figure 10.13).

If Evan has attempted to listen to the quantum channel in order to gain information about photon polarization, and because Evan cannot measure both rectilinear and diagonal polarizations of the same photon, and qubits cannot be copied, cloned or measured, he will alter the photon polarization state and he will cause errors in Bob's measurements, which Alice and Bob will detect during the last steps of the key establishment process and they will abort the key; in this case, they try anew.

The detection of Evan relies on probabilities. At the outset, Alice has selected a string of bits such that they appear with a certain probability p in one state or

Alice's Bit stream	1	0	0	1	0	1	1	1
Alice's random basis	+	X	+	X	+	+	X	+
Polarized photons	→	↗	↑	↘	↑	→	↘	→
Bob's random basis	+	+	+	+	X	+	X	X
Bob's passing polarized photons	→		↑			→	↘	
Public Communi-cation								
Secret key	1		0			1	1	

Figure 10.13. Step-by-step BB84 protocol.

another. Thus, Alice expects to find on the average a certain number of correct bits, and thus a certain key length, during the key establishment process; the expected key length is known as *privacy amplification*. If the channel is tapped by Evan, the statistical behavior will be altered and it will not result the number of correct bits that Alice expects. In this case, Alice retries the key establishment process using reduced privacy amplification; that is, reduce the length of the expected key.

10.3.6 The B92 Protocol

The B92 protocol is a newer version of the BB84 that was developed by Bennett in 1992 [36]. The B92 selects the two dimensional Hilbert space and it is described in terms of any two non-orthogonal basis.

Using the ket notation, the general polarization states of a single photon $|\theta\rangle$ and $|-\theta\rangle$ are used with respect to the vertical, where θ is between 0 and 45 degrees. B92 associates logic "1" symbols with $|\theta\rangle$ and logic symbols "0" with $|-\theta\rangle$.

Based on this, Alice transmits her $\{0,1\}$ bit sequence as a sequence of polarization states $\{|\theta\rangle, |-\theta\rangle\}$. Since the two angles θ and $-\theta$ are not orthogonal, there is no polarization basis and thus no experiment that can distinguish between the two. The details of this protocol are similar to BB84 with some small exceptions described in [36], and thus we will not elaborate on it.

However, since there are only two polarization states and the two angles θ and $-\theta$ remain the same, Evan may try to detect them as follows:

Evan will disrupt the key establishment by trying to measure the angles. Alice and Bob will detect this and they will abort the key. However, as Alice and Bob will repeatedly attempt to establish the key, Evan will have the opportunity to detect the angles.

10.3.7 The K05 Protocol

The K05 protocol was developed as a more general quantum cryptography protocol than BB84 and B92 in an attempt to identify vulnerabilities of realistic quantum networks [37, 38].

The K05 is based on two general subsets of polarization states (Figure 10.14), which are privately defined by Alice.

The K05 protocol uses many polarization states for the binary alphabet {0, 1}, and there is neither known basis (like the BB84) nor two specific states (like the B92) that can be exploited to efficiently eavesdrop. As a consequence, Evan will try in vain to identify all possible polarizations that Alice has determined, particularly if she keeps redefining the two subsets. Figure 10.15 outlines the K05 protocol.

The various steps of the quantum key generation and QKD in K05 is explained in the following logical steps:

1. Alice passes a sequence of binary bits, say 100110111011 … , through a random polarization filter. This sequence is transformed to a sequence of random polarization states. One subset of polarization states is associated with symbol "1" and the other with symbol "0;" the two subsets may be visualized as two regions on the Poincaré sphere. The association of polarization states with logic "1" and "0" are known only to Alice and unknown to anyone else, including Bob.

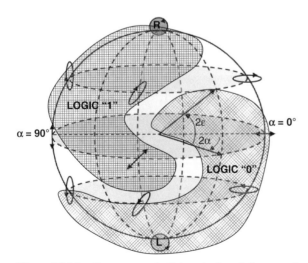

Figure 10.14.   Two areas on the Poincaré sphere define two subsets of polarizations; one area corresponds to logic "0" and the other to logic "1."

Bit sequence:

Alice's Logic sequence:

After passing a polarizing filter:

Bob's polarization states:

  Bob does not know the correct
states. He sends his polarization
sequence to Alice.
Alice tests Bob's sequence and
determines which states were
successful.

Bob's correct states
(as tested by Alice) are:

Alice tells Bob the correct states,
which establish the quantum
(polarization) key:

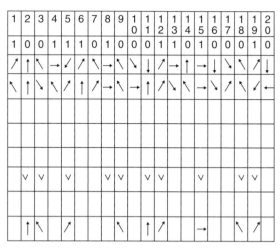

Figure 10.15. The K05 protocol, step-by-step.

2. Bob receives the sequence of polarized photons which he passes through his independently randomly varying polarization filter, but Bob does not know the association between logic value and polarization state.

3. The random polarization states of his filter pass or reject the received randomly polarized photons. That is, a new sequence of logic "1s" and "0s" is generated in which some bits (statistically speaking and over a long string of bits) have the correct logic value that Alice sent, but not all.

4. Assume that Bob's randomly varying polarization filter generates the sequence 010110101001 from the sequence received from Alice. Although this sequence is not what Alice transmitted, the common bits between the two sequences are important here. However, up to this step, neither Alice nor Bob know which bits are common.

5. Bob communicates with Alice over the publicly unsecured channel and he tells Alice the polarization sequence that he used while receiving Alice's polarized photons. However, Bob does not reveal the logic sequence that he generated.

6. Based on Bob's response, Alice performs an experiment. She passes the logic sequence that she sent to Bob through Bob's polarization sequence. Then, Alice compares the initial bit string with the one generated from the experiment and she identifies the bits that are common in the two bit strings.

7. Alice tells Bob which of his filter polarization states in the sequence were used correctly, but without telling him their association with logic "1" and "0." The polarization states that were correctly used constitute the quantum key.

8. When all this is done, Alice encrypts her message with the established key and transmits the encrypted message to Bob, who deciphers it using the same encryption key.

10.3.8 The K08 Protocol

The K08 protocol is an enhancement of the K05, developed by the author S. Kartalopoulos in 2008 [39].

Assume that K08 defines four (2^2) areas on the Poincaré sphere (Figure 10.16) and the alphabet {11, 01, 10, 00}. That is, the PoSs in each area are uniquely associated with one of the four symbols and thus this case corresponds to a quantum quaternary protocol; the alphabet may consists of four unique symbols or of two binary bits (as indicated). As such, with n correct polarization states between Alice and Bob a key length for an equivalent bit string $n2^2$ is established.

The various steps of the quantum key generation and QKD in K05 is explained in the following logical steps:

1. Alice passes a sequence of string of symbols, say |10|01|10|11|10|11| ... , through a random polarization filter; notice that double-bits correspond to each of four possible symbols {11, 01, 10, 00}. This sequence is transformed to a sequence of random, polarization states. One subset of polarization states is associated with symbol "11," another subset with symbol "01," and so on; the four subsets may be visualized as four regions on the Poincaré sphere. The association of polarization states with logic "11," "01," etc. are known only to Alice and unknown to anyone else, including Bob.

2. Bob receives the sequence of polarized photons which he passes through his independently randomly varying polarization filter, but Bob does not know the association between logic value and polarization state.

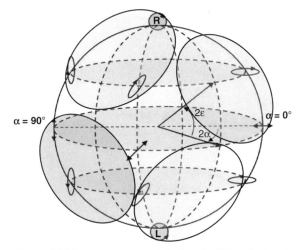

Figure 10.16.   Four areas are defined on the Poincaré sphere to establish a quaternary system.

3. The random polarization states of his filter pass or reject the received randomly polarized photons.

4. Assume that Bob's randomly varying polarization filter generates the sequence |01|11|00|10|00|01| from the sequence received from Alice. Although this sequence is not what Alice transmitted, the common symbols between the two sequences are important here. However, up to this step, neither Alice nor Bob know which bits are common.

5. Bob communicates with Alice over the publicly unsecured channel and he tells Alice the polarization sequence that he used while receiving Alice's polarized photons. However, Bob does not reveal the logic sequence that he generated.

6. Based on Bob's response, Alice performs an experiment. She passes the sequence that she sent to Bob through Bob's polarization sequence. Then, Alice compares the initial symbol string with the one generated from the experiment and she identifies the states that are common in the two symbol strings.

7. Alice tells Bob which of his filter polarization states in the sequence were used correctly, but without telling him their association with symbols "11," "10," and so on. The polarization states that were correctly used constitute the quantum key.

8. When all this is done, Alice encrypts her message with the established key and transmits the encrypted message to Bob, who deciphers it using the same encryption key.

The K08 is now generalized to define an m-ary by subdividing the Poincare sphere in 2^m areas, Figure 10.17 [40]. In this general case, the alphabet consists of

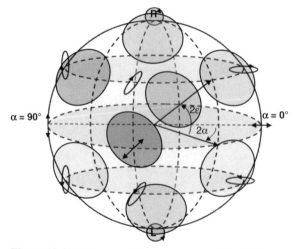

Figure 10.17. In a general quantum scheme, the K08 defines an m-ary system and 2^m areas on the Poincaré sphere.

2^m symbols. Thus, n correct polarization states between Alice and Bob established a key length of n polarizations for an equivalent message $n2^m$ long is established (using binary) or nm long (using symbolic alphabet).

If $m = 6$, then all 26 ASCI characters $\{a, b, c, \ldots\}$, all alphanumerics $\{0, 1, 2, \ldots\}$, and all commonly used symbols $\{!, @, \#, \ldots\}$ can be uniquely represented by $2^6 = 64$ areas. Alternatively, for a fixed length message, the key length can be greatly compressed.

This provides a powerful and more secure protocol because in addition to quantum theoretical principles, other cipher algorithms may also be used.

10.3.9 Entangled Photon Pairs and the Eckert Algorithm

Entangled photons are two photons in close proximity and in orthogonal polarization states [41]. That is, one state of polarization may be linearly vertical and the other linearly horizontal and as the two photons travel together, they interact with each other, although exactly how is not known.

Orthogonal polarized states are created when unpolarized photons propagate through a strongly birefringent crystal, such as the beta-Barium Borate (β-BaB$_2$O$_4$ or BBO). BBO crystals have a trigonal structure, non-linear optical properties, are transparent in the spectral range 189–3500 nm, and exhibit strong birefringence, with refractive indices (extraordinary and ordinary) $n_e = 1.5425$ and $n_o = 1.6551$ @1064 nm, $n_e = 1.5555$ and $n_o = 1.6749$ @532 nm, and $n_e = 1.6146$ and $n_o = 1.7571$ @266 nm. BBO crystals are known to split a high energy photon that propagates through them into two entangled photons, each at half the energy.

Consider that a high energy photon (such as $\lambda = 400$ nm) is directed on a hot BBO crystal. The photon is then split in two photons, each at half the energy ($\lambda = 800$ nm) and with orthogonal polarization states (Figure 10.18). The two photons propagate together keeping their states; they are called entangled photons. Although the two photons can be separated, their entangled states remain.

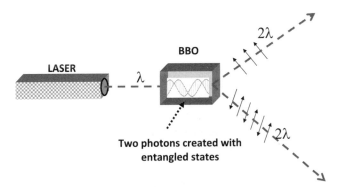

Figure 10.18.   Splitting a high energy photons in two entangled photons with half the energy each.

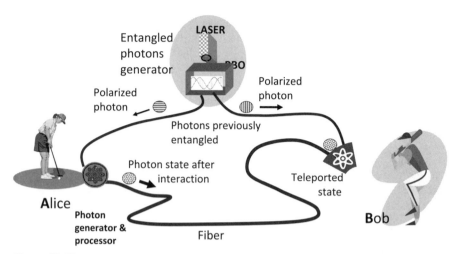

Figure 10.19. A third party creates two entangled photons, one sent to Alice and the other to Bob; Alice and Bob combine it with their one polarized photon in a process to establish a secret quantum key.

The experimental finding by J. Bell in 1964 demonstrated that two individual particles that were previously entangled, when untangled and far from each other exhibited random behavior but also strong correlation; that is, they were still influencing each other. This is known as *Bell state quantum eraser*, and it is not explainable with classical mechanics [42, 43].

The Eckert algorithm is based on entangled photon pairs, which can be made by anyone, Alice, Bob, a third party, or Evan [44]. Entangled photons are distributed so that Alice and Bob end up with one of the two photons but at random polarization (Figure 10.19). Thus, it is impossible for Alice and for Bob to predict what polarization, vertical or horizontal, they will receive. Additionally, if Evan attempts to detect the polarization of any of the two photons, he will destroy it.

10.3.10 Quantum Teleportation

In science fiction movies, teleportation, is depicted as an exciting yet unexplained phenomenon by which matter (including people) disappear from a location and at no time appear elsewhere. This stems from theory that has identified a mysterious quantum-mechanical phenomenon, which Schroedinger coined *entanglement* in a two-part paper that appeared in 1935 and 1936 [45, 46]. These two papers were based on conversations with Einstein, Podolsky, and Rosen who since 1930 were questioning the completeness of quantum mechanics for physical reality [47] and also on a mysterious phenomenon that it is based on *entangled states* of two particles, which Einstein called a *spooky action at a distance*; this currently is known as *Einstein-Podolsky-Rosen correlation* (EPR) phenomenon [48]. The mystery in this phenomenon is that even if the two entangled particles are separated and are at far distances, they still influence the state of each other as if there is a *teleportation of states* [49].

The teleportation effect has been observed in optical experiments with entangled photons and has been deployed in quantum cryptography. According to it, if the state of one of two initially entangled photons—separated and at an arbitrary distance and without communication between—changes, it causes the state of the other to change.

It was John Bell's experiment that observed the photon teleportation phenomenon. Bell used a BBO crystal and a laser that emitted high energy photons. This experiment in a simplified version is described as follows:

An ultraviolet laser emits single (high energy) photons to a hot BBO crystal, which is optically transparent at these frequencies. As a photon travels through the crystal, it is split in two photons, A and B, which now are orthogonally polarized. The two photons exit the BBO crystal and each has half the ultraviolet frequency and energy (this frequency is visible to human eye). The resulting two orthogonally polarized photons are *entangled*.

Subsequent to this, the two entangled photons are separated. The one photon (A) is directed to a photodetector (D_A) through a polarizing filter (P_A), and the other photon (B) is directed to pass through a double slit and then is detected by photodetector (D_B) (Figure 10.20).

When the two photons impinge on their respective photodetectors, one would expect since the two photons by now are completely independent; whatever happens to one photon should not affect the other. Herein lies the mystery: when the polarizing filter P_A allows photon A to pass (because the polarization axis of filter P_A coincides with that of photon A), photon A is detected as well as photon B. However, when the filter P_A is rotated by 90 degrees and it does not allow photon A to pass

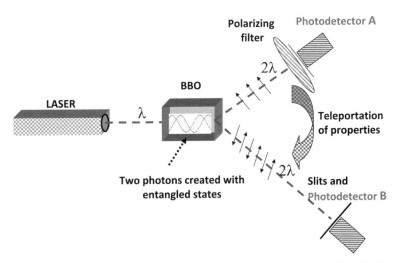

Figure 10.20.   A simplified view of the teleportation experiment [adapted from Bell's experiment].

(because the polarization axis of the polarizing filter P_A is perpendicular to photon's A polarization, known as *polarization erasure*), then not only photon A but also photon B is not detected. That is, in some mysterious way the state of photon A at the polarizer affects the polarization state of photon B even though photons A and B have been separated; that is, the state of photon A is instantaneously teleported to photon B.

In ket notation, the Bell's state is written as:

$$|\Phi^+\rangle = \left(1/\sqrt{2}\right)\left(|0\rangle_A \otimes |0\rangle_B + |1\rangle_A \otimes |1\rangle_B\right)$$

Bell's experiment is a little more complex as each slit has a polarizing substance so that one slit passes counter-clockwise the linear polarization and the other slit clockwise. Thus, the observable quantity is presence or absence of an interference pattern; when photon A passes and is detected, the interference pattern at B is detected, and when photon A does not pass, the interference pattern at B disappears. Thus, the obvious question: How do photons at detector B know of the polarizations state of photons at detector A?

It is this teleportation of states that has been ported in the Eckert's protocol. When teleportation is fully understood and explained, it promises a powerful secure mechanism for the distribution of the quantum cryptographic key in optical communications. In the meantime, several experiments are trying to prove the power of entanglement [50–58].

10.3.11 Vulnerabilities of Pragmatic Quantum Cryptography

To date, several quantum cryptographic algorithms have been developed and have been found to be vulnerable to eavesdropping [59]. Analysis of the weaknesses and vulnerabilities of quantum cryptographic algorithms is used as a tool (such as cryptoanalysis in conventional cryptography) in order to assure the security level before a method is generally deployed. Some current vulnerabilities, inefficiencies and weaknesses of the QC method are [60–62]:

- There are no simple off-the-shelf single-photon sources with controllable single photon rate generation and with controllable photon polarization.
- Optical fiber must maintain the polarization state of photons. However, single mode fiber exhibits variability of geometrical specifications in addition to attenuation, birefringence, dispersion, and other non-linearities that affect the polarization and the propagation properties of photons.
- The fiber link in point-to-point links should remain intact and uniform without splices, connectors and other optical components that change the polarization state.
- A single photon must be perfectly coupled onto optical fiber.
- Optical fiber has absorption and scattering centers: at about 1400 nm, absorption peaks due to OH–, below 1300 nm and above 1620 nm increases due to

absorption and Raleigh scattering. Currently, there is no zero-loss fiber in any part of the useful spectrum. In fact, to overcome this, researchers are thinking of quantum repeaters; that is, subsystems that will receive the polarized signal, restore its strength, and retransmit it. This, of course, may defeat the purpose of QKD because Evan can also have the same subsystem which with minor modifications can receive the signal, copy the polarized key, restore the polarization state of photons and retransmit it to Bob.

- A very long random bit sequence is required to warranty a good encryption key. Because the two filters at each end of the fiber are randomly and independently polarized, the number of bits from Alice's sequence that will pass through Bob's filter are fewer; it is those bits that constitute the encryption key. Thus, in order to warranty a relatively long encryption key (few hundred bits), very long sequences must be used.

- Bob must known when the first bit arrives; that is, single photon synchronization between Alice and Bob is difficult to achieve.

- No acknowledgment by Bob that the negotiated encryption key works reliably or correctly. Bob must know if his polarizing filter behaves as prescribed by Alice from the first arriving photon in the encrypted message. Deciding when the first photon arrives is a task of its own.

- There is no mechanism to confirm that the key has been correctly constructed and that the encrypted message has been correctly received and decrypted. This identifies a potentially serious issue with QC robustness and a lack of verification. What if a malicious attacker affects one or the other polarizing filter? What if a malicious attacker adds propagation delay on the line so that filter synchronization is shifted by a bit period? Will Bob recognize it and reconstruct the message?

- To date, only dedicated point-to-point QC and QKD solutions are contemplated, thus under-utilizing the full bandwidth capacity of WDM links. Only one experimental network contemplates a combined QKD (at 1310 nm) and a channel in the DWDM C-band, thus not utilizing the complete potential of DWDM spectrum.

- An eavesdropper may easily attack the transmitted polarization states on purpose to cause denial of service. So far, the focus in QKD has been to prevent eavesdropping. However, it is equally important to prevent attacks by countermeasure strategies. An attacker may tap the medium and maliciously destroy the QKD process and thus hamper transmission of the encrypted message. In such a case, an eavesdropper is not only a person that needs to "listen" but also one that hinders and deters successful communication between point A and point B; jamming is a well known form of communications deterrence.

- If single-photon bit transmission is not possible, then multi-photon is used from where a small part of the photonic pulse may be extracted from the fiber (using fiber tapping) and thus break the encrypted message (assuming that the sophisticated eavesdropper can also "listen" to the conversation between Alice and Bob).

In conclusion, if quantum cryptography should be applied on practical optical networks [63], Evan, a sophisticated malicious actor who does not want to eavesdrop [64, 65] but to attack the ability of establishing secure communications, may accomplish this by altering the properties of the optical medium with very simple tools, such as by twisting the fiber, stretching the fiber or even pinching the fiber with simple paperclips [66–68]. Doing so, he will:

- Change the propagation and polarization properties of photons in the fiber such that the polarization states transmitted by Alice arrive at Bob in an altered state.

- Bob sends back to Alice an entirely inconsistent polarization pattern.

- Alice, not knowing what Evan has done, tests the received pattern that Bob sent to her, finds the erroneous common bits, establishes the wrong key and sends to Bob a wrongly encrypted message, or she discards the key and starts anew, over and over.

- Bob receives the cipher text but he cannot decipher it successfully, or he starts anew.

- Although Alice and Bob realize that the security of the quantum channel has been compromised, and that secure communication has been disrupted. To continue with secure communication, Alice and Bob must try another fiber, which also may have been compromised.

Although the aforementioned may seem hypothetical, it is a simple, inexpensive and reasonable scenario. This simple experiment has confirmed that altering the state of polarization is very simple; all it takes is a very small tensile or bending force on the fiber, herein called a *"paper-clip attack."*

REFERENCES

1. Quantum Mechanics for Dummies—Electrons Are Weird: http://video.google.com/videoplay?docid=-6109621535224180243
2. Quantum Mechanics—Indirect Observation: http://www.youtube.com/watch?v=Y8IQbL_DnGk
3. MU52—Quantum Mechanics (5 parts): http://video.google.com/videoplay?docid=-6109621535224180243
4. Dr Quantum—Double Slit Experiment: http://www.youtube.com/watch?v=DfPeprQ7oGc
5. Dr Quantum explains The Law of Attraction Scientifically (the double slit) http://video.google.com/videoplay?docid=-8876386367056314090
6. Superposition: http://www.youtube.com/watch?v=qpQABLRCU_0&feature=related
7. Wave Structure of Matter: http://video.google.com/videoplay?docid=-1147049447304385499
8. Dr Quantum explains Entanglement: http://video.google.com/videoplay?docid=-8876386367056314090
9. What Lies Beyond Our Own Space-Time Continuum: http://video.google.com/videoplay?docid=-1147049447304385499
10. D. DEUTCH, "Quantum computational networks", *Proceedings of the Royal Society of London A*, vol. 425, pp. 73–90, 1989.

11. D. DEUTSCH, "Quantum Theory of Probability and Decisions", *Proceedings of the Royal Society A*, vol. 455, pp. 3129–3197, 1999.

12. D. DEUTSCH and P. HAYDEN, "Information Flow in Entangled Quantum Systems", *Proceedings of the Royal Society A*, vol. 456, pp. 1759–1774, 2000.

13. D. DEUTSCH, A. EKERT, and R. LUPPACHINI, "Machines, Logic and Quantum Physics", *Bulletin of Symbolic Logic*, vol. 3, no. 3, September 2000.

14. R. CLEVE, A. EKERT, C. MACCHIAVELLO, and M. MOSCA, "Quantum Algorithms Revisited", *Philosophical Transactions of the Royal Society of London A*, pp. 1–18, 1996, and also in arXiv: quant-ph/9708016, v1, 8 Aug. 1997.

15. R.P. FEYNMAN, R.B. LEIGHTON, and M. SANDS, *The Feynman Lectures on Physics*: Vol. III, Addison-Wesley: Reading, MA, 1965.

16. N. HERBERT, *Quantum Reality: Beyond the New Physics*; Anchor Books: New York, NY, 1985.

17. J.M. BROM and F. RIOUX, "Polarized Light and Quantum Mechanics: An Optical Analog of the Stern-Gerlach Experiment", *Chemical Educator*, vol. 7, 2002, pp. 200.204.

18. S.V. KARTALOPOULOS, *Introduction to DWDM Technology: Data in a Rainbow*, Wiley/IEEE Press, 2000.

19. B. WIELINGA and B.C. SANDERS, "Entangled coherent states with variable weighting", *Journal of Modern Optics*, vol. 40, no. 10, 1993, pp. 1923–1937.

20. B.C. SANDERS, "Entangled coherent states", *Physical Review A*, vol. 45, no. 9, 1992, pp. 6811–6015, and in vol. 46, no. 5, 1992, p. 2966.

21. C.H. BENNETT, G. BRASSARD, S. BREIDBARD, and S. WIESNER, "Quantum Cryptography, or Unforgeable Subway Tokens", *CRYPTO*, 1982, pp. 267–275.

22. C.H. BENNETT and G. BRASSARD, "Quantum Cryptography: Public-Key Distribution and Coin Tossing," *Proceedings of the IEEE International Conference on Computers, Systems, and Signal Processing*, Bangalore, India, 1984, pp. 175–179.

23. C.H. BENNETT and G. BRASSARD, "An Update on Quantum Cryptography", *CRYPTO*, 1984, pp. 475–480.

24. D. DEUTCH, "Quantum computational networks", *Proceedings of the Royal Society of London A*, vol. 425, 1989, pp. 73–90.

25. N. GISIN, G. RIBORDY, W. TITTEL, and H. ZBINDEN, "Quantum cryptography", *Review of Modern Physics*, vol. 74, 2002, pp. 145–195.

26. X. LI, "A quantum key distribution protocol without classical communication", quant-ph/020950, Sept. 6, 2002.

27. E. BIHAM, M. BOYER, G. BRASSARD, J. van de GRAAF, and T. MOR, "Security of Quantum Key Distribution Against All Collective Attacks", LANL e-print quant-ph/9801022.

28. C. FUCHS, N. GISIN, R.B. GRIFFITHS, C.S. NIU, and A. PERES, "Optimal eavesdropping in quantum cryptography. I. Information bound and optimal strategy", *Physics Review A*, vol. 56, 1997. pp. 1163–1172.

29. E. BIHAM and T. MOR, "Security of quantum cryptography against collective attacks", *Physics Review Letters*, vol. 78, 1997, pp. 2256.

30. V.E. BIHAM and T. MOR, "Bounds on information and the security of quantum cryptography", *Physics Review Letters*, vol. 79, 1997, pp. 4034.

31. D. MAYERS, "Quantum key distribution and string oblivious transfer in noisy channel", *Proceedings Of Crypto'96, LNCS 1109*, vol. 343, 1996, pp. 343–357.

32. D. MAYERS, "Unconditional security in quantum cryptography", LANL e-print, quant-ph/9802025.

33. D. DEUTSCH, "Quantum theory, the Church-Turing Principle and the universal quantum computer", *Proceedings of the Royal Society of London A*, vol. 400, 1985, pp. 97.

34. A. PERES, *Quantum Theory: Concepts and Methods*, Kluwer Academic, Dordrecht, 1993.

35. W.K. WOOTTERS and W.H. ZUREK, "A single quantum cannot be cloned", *Nature*, vol. 299, 1982, pp. 802–803.

36. C.H. BENNETT, "Quantum cryptography using any two nonorthogonal states", *Physical Review Letters*, vol. 68, no. 21, 25 May 1992, pp. 3121–3124.

37. S.V. KARTALOPOULOS, "Link-layer Vulnerabilities of Quantum Cryptography", SPIE International Congress on Optics and Optoelectronics, Warsaw, Poland, 8/28/05-9/2/2005, *Proceedings of SPIE*, vol. 5954, pp. 5954OH-1 to 5954OH-7.

38. S.V. KARTALOPOULOS, "Identifying vulnerabilities of quantum cryptography in secure optical data transport", Unclassified Proceedings of Milcom 2005, 10/17-20/05, Atlantic City, session: Comm. Security I, invited paper # 678, on CD-ROM, ISBN # 0-7803-9394-5.

39. S.V. KARTALOPOULOS, "K08: A general Quantum Protocol", submitted for publication.

40. S.V. KARTALOPOULOS, "A generalized Quantum protocol with m-ary key dimensionality", submitted for publication

41. A. POPPE, et al., "Practical quantum key distribution with polarization entangled photons", *Optics Express*, vol. 12, no. 16, 2004, pp. 3865–3871.

42. J.S. BELL, "On the Einstein-Podolsy-Rosen paradox", *Physics*, vol. 1, pp. 195–200, 1964. Reprinted in J.S. Bell, *Speakable and Unspeakable in Quantum Mechanics*, Cambridge University Press, Cambridge, 1987.

43. A. ASPECT et al., 'Experimental Tests of Bell's Inequalities Using Time-Varying Analyzers', *Physics Review Letters*, vol. 49, 1982, pp. 1804–7.

44. A.K. EKERT, Quantum cryptography based on Bell's theorem. *Physics Review Letters*, vol. 67, no. 6, 1991, pp. 661–663.

45. E. SCHROEDINGER, "Discussion of Probability Relations Between Separated Systems", *Proceedings of the Cambridge Philosophical Society*, vol. 31, 1935, pp. 555–563.

46. E. SCHROEDINGER, "Discussion of Probability Relations Between Separated Systems", *Proceedings of the Cambridge Philosophical Society*, vol. 32, 1936, pp. 446–451.

47. A. EINSTEIN, B. PODOLSKY, and N. ROSEN, "Can quantum-mechanical description of physical reality be considered complete?" *Physical Review*, vol. 47, 1935, pp. 777–780. Reprinted in *Quantum theory and measurement* (J. A. Wheeler and W. Z. Zurek, eds.), Princeton University Press, 1983.

48. D. DIEKS, "Communication by EPR devices", *Physics Letters A*, vol. 92, no. 6, 1982, pp. 271–272.

49. MARK BUCHANAN, "Quantum Teleportation", *New Scientist*, 14 March 1998.

50. G. RIGOLIN, "Quantum Teleportation of an Arbitrary Two Qubit State and its Relation to Multipartite Entanglement", *Physics Review A*, vol. 71, no. 032303, 2005.

51. L. VAIDMAN, "Teleportation of Quantum States", *Physics Review A*, 1994.

52. D. BOUWMEESTER, J.-W. PAN, K. MATTLE, M. EIBL, H. WEINFURTER, and A. ZEILINGER, "Experimental Quantum Teleportation", *Nature*, v. 390, 1997, pp. 575–579.

53. D. BOSCHI, S. BRANCA, F. DE MARTINI, L. HARDY, and S. POPESCU, "Experimental Realization of Teleporting an Unknown Pure Quantum State via Dual classical and Einstein-Podolsky-Rosen channels", *Physics Review Letters*. Vol. 80, no. 6, 1998, pp. 1121–1125.

54. I. MARCIKIC, H. de RIEDMATTEN, W. TITTEL, H. ZBINDEN, and N. GISIN, "Long-Distance Teleportation of Qubits at Telecommunication Wavelengths", *Nature*, vol. 421, 2003, pp. 509.

55. R. URSIN et al., "Quantum Teleportation Link across the Danube", *Nature* vol. 430, 2004, pp. 849.

56. L. VAIDMAN, "Using teleportation to measure nonlocal variables", quant-ph/0111124.

57. C.H. BENNETT, G. BRASSARD, C. CR'EPEAU, R. JOZSA, A. PERES, and W. WOOTTERS, Teleporting an unknown quantum state via dual classical and EPR channels. *Phys. Rev. Lett.*, 70:1895–1899, 1993.

58. C.H. BENNETT, D.P. DIVINCENZO, J.A. SMOLIN, and W.K. WOOTTERS, Mixed state entanglement and quantum error correction. *Phys. Rev. A*, 54:3824, 1996. arXive eprint quant-ph/9604024.

59. D.R. KUHN, "Vulnerabilities in Quantum Key Distribution Protocols", quant-ph/0305076, May 12, 2003.

60. S.V. KARTALOPOULOS, "Link-layer Vulnerabilities of Quantum Cryptography", SPIE International Congress on Optics and Optoelectronics, Warsaw, Poland, 8/28/05-9/2/2005, *Proceedings of SPIE*, vol. 5954, pp. 5954OH-1 to 5954OH-7.

61. S.V. KARTALOPOULOS, "Identifying vulnerabilities of quantum cryptography in secure optical data transport", *Unclassified Proceedings of Milcom 2005*, 10/17-20/05, Atlantic City, session: Comm. Security I, invited paper # 678, on CD-ROM, ISBN # 0-7803-9394-5.

62. S.V. KARTALOPOULOS, "Is optical quantum cryptography the 'Holy Grail' of secure communication?" *SPIE Newsroom Magazine*, April 2006, http://newsroom.spie.org/x2260.xml?highlight=x537.

63. S.V. KARTALOPOULOS, *DWDM: Networks, Devices and Technology*, Wiley/IEEE Press, 2003.

64. M.J. WERNER and G.J. MILBURN, "Eavesdropping using quantum-nondemolition measurements", *Physical Review A*, vol. 47, no. 1, January 1993, pp. 639–641.

65. S.M. BARNETT, B. HUTTNER, and S.J.D. PHOENIX, "Eavesdropping strategies and rejected-data protocols in quantum cryptography", *Journal of Modern Optics*, vol. 40, no. 12, December 1993, pp. 2501–2513.

66. S.V. KARTALOPOULOS, *Next Generation Intelligent Optical Networks: From Access to Backbone*, Chapter 10, Springer, 2008.

67. S.V. KARTALOPOULOS, "Is optical quantum cryptography the 'Holy Grail' of secure communication?" SPIE Newsroom Magazine, April 2006, http://newsroom.spie.org/x2260.xml?highlight=x537

68. S.V. KARTALOPOULOS, "Quantum Cryptography for Secure Optical Networks", *Proceedings of the ICC 2007 Conference, Computer and Communications Network Security Symposium*, June 24–28, 2007, ISBN 1-4244-0353-7 (on CD-ROM).

Chapter 11

Next Generation Optical Network Security

11.1 INTRODUCTION

Chapters 7 through 9 presented several types of networks, wireless, wired and optical, their security features and also their security issues.

This chapter focuses on the next generation optical network, which although SONET/SDH-based, has certain new standardized protocols in place to provide the flexibility to the synchronous network to cost-efficiently transport all types of payloads to voice, video, data, Internet, and so on. These protocols may also be exploited to enhance information security across the network. However, at the access points, information security still relies on end-to-end cryptographic systems as already discussed [1–7].

Notice that although the next generation network is fiber-optic, it is doubtful that quantum cryptography and teleportation can be used across the network for the simple reason that quantum security methods are most suitable at the link layer. The next generation optical network consists of links and nodes [8, 9].

Links do not consist of perfect fiber; fiber has birefringence due to non-uniform refractive index ($B \sim |n_1 - n_2|$), and due to non-uniform geometry (the core does not remain perfectly circular but according to ITU-T, its diameter may vary within $\pm 0.5\,\mu m$ along its length). For example, if the fiber attenuation constant is $\sim 0.2\,dB/km$, the overall attenuation for a 100 km link is 20 dB. Clearly, in single photon transmission this is easily translated to a high percentage probability loss; that is, many photons will never make it to the photodetector. Because this photon loss is random, Bob cannot know that photons did not arrive because of attenuation or because his receiver is not well synchronized. Moreover, links with length that exceed 60–80 Km, most likely need amplification (either EDFA or Raman). In this case, there are two issues:

- it is doubtful that the quantized state of photons can be maintained.
- it is doubtful that single photon can be used, as amplifier spontaneously emitted noise (ASE) will bury the single photon in ASE photonic noise.

Security of Information and Communication Networks, by Stamatios V. Kartalopoulos
Copyright © 2009 Institute of Electrical and Electronics Engineers

Nodes will be all-optical transparent but some will be opaque.

- All-transparent nodes, depending on design, may be suitable to quantum security, for many all-transparent nodes will be very difficult to maintain the quantized state of photons as component insertion loss and component non-linearity will either destroy photons or alter their characteristics; such optical and photonic components are amplifiers, equalizers, dispersion compensators, filters, etc., on the nodes and on the links that the quantum state of photons will be impossible to maintain [10–12].

- Opaque nodes convert optical information to electrical and back to optical. Consequently, the quantized state of photon is destroyed at the photodetector, something that even can exploited very easily. Opaque nodes are needed after the optical signal has traversed 6–7 optical amplifiers; currently, commercial optical amplifiers perform signal amplification and they do not restore pulse shape or retiming; although single photon transmission does not need the last two, the overall commercial DWDM backbone network must be designed to include these two functionalities. This is one more argument that the quantum network cannot use the all-optical commercial network but a private fiber network.

Nevertheless, there are certain aspects of the next generation optical network that can be conveniently explored to enhance link and network security, which is the objective of this chapter.

11.2   STANDARDIZED PROTOCOLS FOR COST-EFFICIENT OPTICAL NETWORKS

The initial synchronous optical network (SONET) in the U.S., and synchronous digital hierarchy (SDH) in Europe [13], were two similar standardized protocols developed in the 1980s for efficient transport of synchronous payloads.

However, the 1990s witnessed an explosion of data, and their transport over the well-established and very robust SONET/SDH network was not cost-efficient because data protocols such as the Internet, point-to-point, and others were not cost-efficiently mapped onto virtual tributaries or unit (these are predefined containers in the SONET/SDH protocols). In addition, the SONET/SDH protocol was defined for a single wavelength (or optical channel) per fiber, 1310 nm for long fiber lengths and 1550 nm for medium lengths. In the 1990s, DWDM technology emerged, which multiplexed hundreds of wavelengths (channels) in a single fiber. However, SONET/SDH does not considered channel number, which is necessary for traffic, maintenance and protection.

11.2.1   The Generic Framing Protocol

The Generic Framing Procedure (GFP) defines a flexible encapsulation framework for traffic adaptation of broadband transport applications [14–17]. The GFP supports client control functions allowing different client types to share a channel. The GFP efficiently maps broadband data protocols onto multiple concatenated STS-1 payloads in a revised SONET/SDH frame. It is the GFP, along with the revised SONET/SDH frame, which constitute the Next Generation SONET/SDH (NGS), which is differentiated from the legacy SONET/SDH.

Some of the GFP over NGS features are:

- The GFP maps a physical or logical layer signal to a byte synchronous channel.
- The NGS supports various network topologies.
- The GFP exhibits low-latency of packet-oriented or block-coded data streams.
- The GFP supports differentiated QoS meeting service level agreement (SLA) requirements.
- The NGS supports a flexible encapsulation of diverse protocols onto GFP generalized frames which are mapped onto the SPE of the revised SONET/SDH.
- GFP with NGS has enhanced features that improve bandwidth utilization.
- GFP defines different frame lengths as well as different client-specific frame types for payload and for management. Therefore, GFP supports time multiplexing of different frames, on a frame by frame basis, to increase transmission efficiency.
- GFP provides a continuous stream of frames. At the absence of client frames, GFP inserts idle frames.
- NGS is an integrated and interoperable transport platform that supports cost-efficiently packet and circuit switching services.
- GFP with NGS is suitable for wavelength division multiplexing (WDM), short reach, intermediate reach or long reach (SR, IR, LR) in which each wavelength caries the equivalent of hundreds of client signals. The synergy of GFP, NGS, and WDM constitutes the NGS over WDM.

In conclusion, GFP with the Next Generation SONET/SDH over WDM has improved bandwidth utilization for all types of traffic—both synchronous and asynchronous—and a transport capability at Tbps per fiber. To understand this, one needs to examine how the GFP protocol works.

11.2.1.1 *GFP Client Payload Multiplexing*

GFP defines a payload area with its own control fields that include linear and ring extensions. Thus, payloads of the same type, such as GbE, and from different clients may be multiplexed using either a sequential round-robin method (as in DS1 and DS3) or after a well-established queuing schedule. The multiplexing method depends on size of client frames and on frequency of arrival.

- Round-robin is used when client frames to be multiplexed are (almost) synchronous and preferably of the same length.
- A queuing schedule is used when client frames are asynchronous, there is substantial variability in frame length, and there is a random arrival of packets (or a random inter-packet separation). In this case, the incoming client frames is buffered and retimed to reduce jitter, and then multiplexed.

11.2.1.2 GFP Frame Structure

The GFP platform defines a flexible frame structure (Figure 11.1). The GFP frame consists of a "core header" and a "payload area."

- The "core header" supports non-client specific data link management functions such as delineation and payload length indication (PLI) and core header error control (cHEC).
- The "payload-area" supports client-specific client-to-client connectivity and error control. It consists of three fields:
  - The "client information data" that may have a fixed or variable length bounded by two fields,
  - The "payload header" field that contains specific information that pertains to payload type and error control, and
  - The "payload frame sequence" (FCS) field which is used for payload error control.

Error control is segregated between the GFP adaptation layer and user data. This allows for sending to the intended receiver corrupted frames; in video and audio applications, this is preferred because corrupted frames are better than no frames at all. It also allows for end-to-end user-defined error-correction coding.

GFP defines two classes of functions, common and client-specific:

- ***Common*** are PDU delineation, data link synchronization, scrambling, client PDU multiplexing and client-independent performance monitoring.

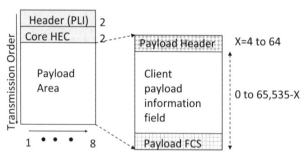

PLI=Payload Length Indicator
MSB=Most significant byte
LSB=Least significant byte
FCS=Frame check sequence
HEC=Head error control

Figure 11.1.   Payload of the GFP frame.

- *Client-specific* are mapping client PDU onto the GFP payload, and client specific OA&M (operations, administration, and maintenance).

It also distinguishes two types of client frames, one for client data frame (CDF), and the other for client management frames (CMF). A CMF transports information related to GFP connection or client signal management. A CMF allows clients to control the client-to-client connection.

11.2.1.3 *GFP Header Error Control and Delineation*

The core header consists of four bytes only. The first two define the length of the payload and the other two a standard cyclic redundancy check code (CRC-16) that protects the integrity of the core header from errored bits.

The core header error control (cHEC) mechanism identifies multiple errors and correct single errors. At the transmitting end, where the GFP is formatted, the 16-bit CRC code is calculated over the core header and is stored in the third and fourth byte of the GFP core header. At the receiving end, the incoming header is calculated in real time, detecting possible multiple errors and correcting single errors.

The delineation function identifies the start of a GFP frame, known as delineation. Delineation is accomplished with the CRC-16 in the two core-HEC bytes. GFP frame delineation using the cHEC is a process similar to that used in the asynchronous transport mode (ATM). According to it, the payload length and the error-control check (cHEC) is all the information needed for delineation. GFP delineation requires that there are two consecutive cHEC matches. Whenever a cHEC match fails (a non-correctable mismatch), then the state machine moves to the re-synchronization state.

A field in the GFP frame header contains the payload length. This explicit indicator may change from frame to frame to allow for variable length PDUs, or it may remain constant to support TDM-like synchronous services.

11.2.1.4 *GFP Frame Structure*

The payload type is indicated by the binary value of a three-bit field, the payload type indicator (PTI), which is in the payload header. Currently, two values are defined, one for user data frames and another for client management frames (PTI = 100). The type of payload mapped in the GFP payload area is indicated in an eight-bit byte, the user payload type (UPI). The following payload types are defined:

0×01: Frame-mapped Ethernet

0×02: Frame-mapped PPP (including IP and MPLS)

0×03: Transparent-mapped Fibre channel

0×04: Transparent-mapped FICON

0×05: Transparent-mapped ESCON

0×06: Transparent-mapped GbE

0×07: Reserved

0×08: Frame-mapped multiple access protocol over SDH

0×09–$0 \times$ FE: Reserved

0×00 and $0 \times$ FF: unavailable

11.2.1.5 GFP Modes

GFP specifically allows transport modes within the same transport channel. This is accomplished by defining two modes: the Frame-Mapped GFP (GFP-F), and the Transparent-Mapped GFP (GFP-T).

The GFP-F mode considers a variable frame length and is optimized for packet switching applications, such as IP, native point-to-point protocol (PPP), Ethernet (including GbE and 10 GbE), and multi-protocol label switching (MPLS) traffic. A variable frame assumes buffering and MAC hardware that implies latency but also the ability to recognize and discard idle packets. The GFP-F features are:

- From Layer-2 or higher into GFP frames one at a time.
- It uses HEC for delineation.
- It supports rate adaptation and multiplexing at packet level.
- It aggregates the frames at the STS and VT level.
- It supports most packet data types.
- It requires MAC awareness.
- It requires buffering.
- It introduces latency.
- It supports client data frames (CDF) for both client data and management. CDF consists of a 4-byte core header (CH), and 0–65,535 payload area (PA).
- It supports control frames (for idle, and for OA&M).

The GFP-T mode considers a fixed frame length and is optimized for applications that require bandwidth efficiency and delay-sensitivity; thus, it requires no buffering. Applications include Fibre channel (FC), FICON, ESCON, and Storage area networks (SAN). The GFP-T features are:

- N to 1 mapping of client packets.
- It operates on each arrived byte.
- No latency is introduced.
- No MAC is required. Only 8 B/10 B coding.
- The PHY (physical) layer is terminated.
- Because there is no buffering, it retains idle frames.

11.2.2 The Link Capacity Adjustment Scheme LCAS

The *Link Capacity Adjustment Scheme* (LCAS) allows containers in the NGS to be dynamically added or removed in order to meet user bandwidth requirements [18].

LCAS is accomplished using control packets to configure the path between source and destination. The control packet is transported over the H4 byte for high-order virtual containers (HO VC) in a super-frame, and over the K4 byte for low-order virtual containers (LO VC). A superframe consists of N multiframes; each multiframe consisting of 16 frames. In the LCAS example (Figure 11.2), STS-1-2 may be deemed unnecessary and thus it may be released to transport other traffic.

In LCAS, from source to destination path, the H4 overhead byte contains the following control information in the 4-bit nibble (bit 5 to bit 8):

- Fixed (0000): no LCAS mode
- Add (0001): this member to be added
- Norm (0010): no change, steady state
- Idle (0101): no part of group to be removed
- CRC-8 (0110) & (0111): control packet protection
- RS-Ack (1010): from source to destination, requested changes are accepted
- Reserved (1011)
- Reserved (1100)
- Reserved (1101)

Other control information over the H4 byte in the super-frame is:

- GID: Group ID; members in a group have same ID
- EOS: end of sequence

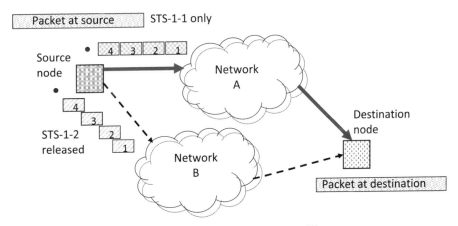

Figure 11.2. The LCAS protocol optimizes frame transport over different routes.

- DNU: Do not use, failure detected
- MST: Member status from source to destination, either OK or failed

11.2.3 Concatenation Efficiency

The legacy SONET hierarchy provides bandwidth granularity of 1.5, 2, 45, or 155 Mbps. With virtual concatenation, packets of data are mapped over several concatenated Next Generation SONET/SDH frames with the result of increasing bandwidth efficiency. Concatenation may be summarized as follows:

- *Containers* are the building blocks of NG-SONET/SDH.
- *Contiguous concatenation* is the method of using multiple contiguous containers in the same synchronous payload envelope to make a large container.
- *Virtual concatenation* is the method of using multiple containers in different synchronous payload envelopes to make a large container. However, each SPE transports the containers across independent routes in the network.

Table 11.1 provides a comparison between contiguous and virtual concatenation and Table 11.2 provides the efficiency for contiguous concatenation (CC) and for virtual concatenation (VC).

Accordingly, NGS offers a complete solution for all payload types, an efficiency that may reach 95% (ATM is only 80%), payload distribution that is as flexible as the data network, and a time-tested protocol (SONET/SDH) that is applicable to long-haul, wide-area and metro-area networks. In addition, NGS has inherited a fast switching to protection (less than 50 msec) and is applicable to WDM. Thus, using LCAS over GFP over NGS over WDM, one has a powerful way of cost-efficiently transporting any type of payload with the expected quality of service and latency optimization.

Table 11.1.   Comparison between contiguous and virtual concatenation

Contiguous concatenation	Virtual concatenation
Limited granularity of VC size	More flexible Low order and High order granularity
Contiguous containers travel over the same path	Containers travel over separate paths
All network element on the path must be aware of CC	Only the end network element must be aware of VC
Transparent to net. management	Requires network management
No differential delay	Potential differential delay
No sequential numbering for alignment	Requires sequential numbering for alignment

Table 11.2. Efficiency of CC and VC for various payloads

Service	Data Rate	CC			VC		
		SONET	SDH	Efficiency	SONET	SDH	Efficiency
Ethernet	10 Mbps	STS-1	VC-3	20%	VT-1.5-7v	VC-12-5v	90%
Fast Ethernet	100 Mbps	STS-3c	VC-4	67%	STS-1-2v	VC-3-2v	100%
GbE	1 Gbps	STS-48c	VC-4-16	42%	STS-1-21v	VC-4-7v	95%
ESCON	200 Mbps	STS-12c	VC-4-16	33%	STS-1-4v	VC-3-4v	100%
FC	1 Gbps	STS-21c		85%	STS-1-18v		95%
ATM	25 Mbps	STS-1	VC-3	50%	VT-1.5-16v	VC-12-12v	98%

As a consequence, several data equipment designers and data networking companies have found that IP packets over NGS (PoS) is a viable and cost-efficient solution for backbone networking. Despite this, there are some that believe that it is more efficient to map IP directly onto the optical channel, implying that all payload types should be over IP (voice over IP, video over IP, etc.). Thus, there are two schools of thought when one considers WDM technology, the NGS-centric and the IP-centric. Which one will prevail? Obviously the one that offers better security, data and network, wins.

11.3 SECURITY IN THE NEXT GENERATION SONET/SDH

The Next Generation SONET/SDH over WDM, using GFP and LCAS protocols, provides a standardized, robust, and efficient method that transports all types of payload, asynchronous data and synchronous traditional TDM.

The Generic Framing Procedure (GFP) accomplishes the necessary encapsulation of GbE, IP, FC, and others by mapping in virtual concatenated NGS frames.

Finally, using LCAS an improved bandwidth utilization and efficiency is achieved.

According to the general scheme of NGS, synchronous traffic such as DS-n/E-n is mapped in SONET/SDH (VT/TU groups), data (IP, Ethernet, FC, PPP) is encapsulated in GFP (first in GFP client-dependent and then in GFP common to all payloads), and then mapped into the payload envelope of the NGS STS-n frames. And as this adaptation to GFP and mapping in NGS takes place, overhead bytes and pointers are formed to construct a NG-S frame. Figure 11.3 depicts the logical major steps to map client traffic over NG-S and Figure 11.4 shows the mapping over NG-S with LCAS.

Based on this, the GFP and LCAS may also be used as a means for network security.

For example:

A client's payload may be mapped in random VTs specified by the secret key. In addition, using dynamic frame allocation over different paths LCAS adds another

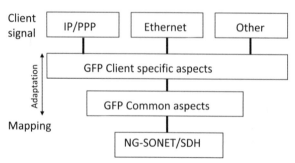

Figure 11.3. Mapping client traffic over NGS.

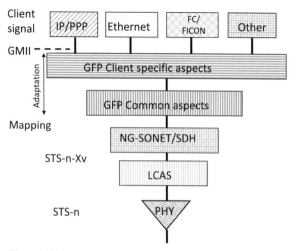

Figure 11.4. Mapping client traffic over NGS including LCAS.

complexity. Thus, using end-to-end encryption, dynamic mapping with GFP, dynamic bandwidth allocation and path selection with LCAS, dynamic WDM channel selection, and intrusion detection with countermeasures, increased information and network security is achieved.

In conclusion, NGS has inherent network security features that its predecessor SONET/SDH did not have, yet it transports all types of payload with the expected QoS, data integrity and latency commensurate to the payload type.

REFERENCES

1. *Next Generation SONET/SDH: Voice and Data*, IEEE/Wiley, 2004.
2. ANDRES SIERRA, NIJAS KOLOTHODY, and STAMATIOS V. KARTALOPOULOS, "Next Generation SONET/SDH with Enhanced Service Protection", *WSEAS Transactions on Communications*, Iss. 9, vol. 4, Sept 2005, pp. 888–898.

3. A. SIERRA, N. KOLOTHODY, and S.V. KARTALOPOULOS, "LCAS and GFP for Service Protection in Next Generation SONET/SDH", *Proceedings of the 9th WSEAS CSCC Multiconference: Communications'05*, Greece, July/14–16/2005, paper 497-080 on CD-ROM.

4. ALAN HARRIS, ANDRES SIERRA, S.V. KARTALOPOULOS, and J. SLUSS, Jr., "Security Enhancements in Novel Passive Optical Networks", ICC 2007 Computer and Communications Network Security Symposium, 6/24–28/07, ISBN 1-4244-0353-7 (CD-ROM).

5. S.V. KARTALOPOULOS and ANDRES SIERRA, "Engineering a Scalable and Bandwidth Elastic Next Generation PON", OFC/NFOEC 2007, March 25–29, 2007, Anaheim, CA, paper NThB4, on CD-ROM, ISBN 1-55752-830-6.

6. N. KOLOTHODY, A. SIERRA, and S.V. KARTALOPOULOS, "Network Protection on Next Generation PON", NOC 2005, 10th European Conference on Networks and Optical Communications, University College London, July 5–7, 2005, pp. 107–114.

7. S.V. KARTALOPOULOS, "Next generation WDM networks with multiple channel protection and channel equalization", IEEE 2005 Workshop on High Performance Switching and Routing (HPSR2005), May 12–14, 2005, Hong Kong, China, Session 4C: Resilience, Protection and Restoration II, On CD-ROM, paper no. 1568951324.

8. S.V. KARTALOPOULOS, *Next Generation Intelligent Networks: From Access to Backbone*, Springer, 2008.

9. S.V. KARTALOPOULOS, *DWDM: Networks, Devices and Technology*, IEEE/Wiley, 2003.

10. S.V. KARTALOPOULOS, *Introduction to DWDM Technology*, IEEE/Wiley, 2001.

11. S.V. KARTALOPOULOS, *Fault Detectability of DWDM Systems*, IEEE/Wiley, 2001.

12. S.V. KARTALOPOULOS, "DWDM: Shaping the future Communications Network", IEEE Potentials Magazine, October/November 2004, pp. 16–19.

13. S.V. KARTALOPOULOS, *Understanding SONET/SDH and ATM*, IEEE/Wiley, 1999; also, Prentice Hall of India, 2000.

14. ITU-T Recommendation G.7041/Y.1303, "The Generic Framing Procedure (GFP) Framed and Transparent", December 2001.

15. S.S. GORSHE and T. WILSON, "Transparent Generic Framing Procedure (GFP): A Protocol for Efficient Transport of Block-Coded Data through SONET/SDH Networks", *IEEE Communications Mag.*, vol. 40, no. 5, May 2002, 88–95.

16. E. HERNANDEZ-VALENCIA, M. SCHOLTEN, and Z. ZHU, "The Generic framing procedure (GFP): An Overview", *IEEE Communications Mag.*, vol. 40, no. 5, May 2002, pp. 63–71.

17. E. HERNANDEZ-VALENCIA, "Generic framing procedure (GFP): A next-generation transport protocol for high-speed data networks", *Opt. Networks Mag.*, vol. 4, no. 1, Jan/Feb 2003, pp. 59–69.

18. ITU-T Recommendation G.7042/Y.1305, "Link Capacity Adjustment Scheme (LCAS) for Virtual Concatenated Signals", Nov. 2001.

Chapter 12

Biometrics in Communication Networks

12.1 INTRODUCTION

Biometrics and biometry is a technology that combines biology, natural individualistic markings, shapes, and sounds as well as idiosyncratic behavior with engineering and statistical methods in an effort to uniquely identify and authenticate a person or an animal. In a general sense, biometrics and biometry employs mathematical and statistical methods to analyze data gathered from mammals and agricultural crops to identify statistical singularities that help to diagnose and solve problems in biological sciences. In a specific sense, biometrics is used in physical access, ID confirmation or verification, surveillance, logical access (or data access), financial transactions, attendance and voting, healthcare and hospitals, agriculture, marine sciences, and more.

For example, a person's facial characteristics (shape of face, distance between eyes and tip of nose, ears, etc.) can be used to identify that person. When meeting a friend, one can tell who the person is even if he/she resembles others. There are certain characteristics that can serve as a set of identification data, and in some cases, it may be a combination of sets of data that is unique about each person. The same is true for animals; zebras have unique black/white line patterns, and whales have unique markings at the tip of their tail.

In addition to markings, sounds are also examined. Upon hearing someone's voice, one can recognize him/her. Likewise one can hear someone walking, and without having seen the person, tell who the person is.

As a result of increased ID fraud, biometric authentication methods have been advanced to include features extracted from fingertips, palm, iris, writing, typing on a keyboard, and DNA, based on standardized methodologies [1, 2].

Biometrics is not a new topic. It has been previously addressed with rather rudimentary methods such as a photograph on a passport, a fingerprint and a signature on a document. These methods, however, have been repeatedly defeated by impostors. Recent movies show how easy it is, with technological advancements, to

Security of Information and Communication Networks, by Stamatios V. Kartalopoulos
Copyright © 2009 Institute of Electrical and Electronics Engineers

steal one's fingerprint, mimic one's identification card in order to gain unauthorized access, and alter one's voice to sound like someone else's. Therefore, biometrics is looking beyond the characteristics that can be easily altered with cosmetic surgery or even with other simpler methods; describing this is beyond this book's objective.

However, the particular methodology, the mathematical algorithms to form templates, and the accuracy and performance of each methodology is not the main problem here. The problem is Evan, a master of disguise and a malevolent impostor who pretends to be someone else for his own profit. In fact, Evan may temporarily or even permanently change the characteristics of his face, voice, appearance and other features to accomplish his dubious plans.

12.2 BIOMETRIC TYPES

Biometry extracts and classifies features that uniquely identify individuality. Therefore, biometrics should be measurable. Typically, a biometric apparatus extracts vectorial data, which are digitized and sent to a computer for further processing. The computer, according to a program forms templates with features that constitute the biological signature of a person, collectively known as "biometrics." Features can be physiological and/or behavioral.

Physiological are:

- Iris of the eye
- Retina of the eye
- Fingertips (minutiae, forks, orientation, etc.)
- Palm (size, shape, finger orientation, lines, etc.)
- Face (size, shape, eyes, nose, lips, etc.)
- Dental x-rays
- Body shape and height
- Voice (spectral distribution, etc.)
- DNA

Behavioral are:

- Signature
- Writing style
- Syntax, vocabulary, phraseology, etc.
- Computer keystroke patterns
- Speaking
- Walking
- And more.

Physiological or behavioral data are initially obtained from end devices, such as an automated fingerprint identification system (AFIS) or an automated iris identification system (AIIS), are organized in templates, and are stored in a data base for later use. These templates are compared with newly extracted templates to authenticate a person's ID if a match is found.

- Patterns of fingerprint minutiae, ridges and other features are unique to a person. However, fingerprinting has been shown not to be foolproof as the fingerprint of a person may be lifted from a drinking-glass and then used by an impostor.
- The iris minutiae of a person's eye are unique. However, iris identification devices may also be fooled.
- The voice frequency spectrum is a characteristic for each person. However, recent technology has shown that with proper frequency synthesizers, one's voice can be modified to emulate another person's voice.
- The signature of a person has been used as an identifier for centuries. However, a large number of forged signatures have shown that the signature itself is a weak method to prove one's identity.
- Certain facial and anatomical characteristics may be unique to a person. However, it has been shown that they, too, can be temporarily or permanently altered to emulate someone else's characteristics.
- Habitual walking patterns may be used to identify a person. However, good imitators can mimic one's walking patterns.
- Typing patterns may be used to identify a person. This method may not be robust as injuries and age may alter the same person's typing characteristics.
- DNA is a new method that uniquely identifies a person's identity. However, current DNA methods require sophisticated tools and thus DNA has not been deployed as widely yet as an on-the-spot identification method. Nevertheless, nanotechnology is expected to make DNA miniaturized devices readily available. Even so, DNA may easily be stolen and used by unauthorized persons. Cloning techniques may also work against DNA ID as molecules with same or similar DNA may be reproduced to fool the biometrics authenticator.

In conclusion, it seems that a more robust and safer method for personal identity verification [3, 4] should not use a single biometric type but a combination or fusion of biometrics types (static and temporal), such as the fingerprint and face and iris, voice and face and iris, DNA and fingerprint, and so on. The more complex and comprehensive the biometric system is, the more difficult it is to compromise for the impostor. However, the performance of the biometric system, that is, how fast and successful is to identify correct matches and mismatches, is equally important but outside the scope of this book. However, a well-known and quick way to compare pre-stored with newly extracted templates has been content-addressable memory [5].

12.3 BIOMETRICS AND CRYPTOGRAPHY

Cryptography is the art of altering text or digital data so that they become meaningless to a third party and meaningful only to the rightful recipient. Conversely, cryptanalysis is the art of capturing cryptographic data and based on mathematics, statistical analysis and other methods, break the code and discover the key(s). If this is accomplished, then data altering, data copying, data moving from one location to another, data injecting, and more is possible. Therefore, biometric data should be encrypted. In addition, the database system should be physically secured and tamper-proof to deter unauthorized entry into the system and protect data from electronic copying, cloning, and destroying.

12.4 LOCAL AND REMOTE AUTHENTICATION

In general, biometric devices are considered end devices of an access communications network, regardless of how large or small the network is. The applicability of biometrics ranges from gaining entrance to a building or home, gaining easy access to avoid long security lines at airports, operating a machine (car, computer), accessing bank accounts, gaining access to communications networks, and so on.

For biometrics to perform authentication, one or more biometric templates of the subject needs to be formed and stored in the database. Thereafter, authentication can be performed. Because biometric templates (or data) may be to may be intercepted by a third party during transfer to the database, data needs to be encrypted.

The authentication process may be local (open the door of a car, home, machine) or it may be remote (access to a corporate building and its sections, access a home supervised by a security entity, access banking accounts, ATM machines, and so on) [6, 7].

- Local biometrics authentication, also known as a closed loop biometric system, is accomplished by previously storing fingertip characteristics or the image of the iris in local memory. When the operator needs to start the machine, he/she sweeps a designated fingertip or has his/her iris scanned, and the machine compares the obtained data with the pre-stored data. If there is a match, authorization is granted. If not, it does not start. Because this method does not require a communications network, it is not of interest to this book.

- Remote authentication considers biometric terminals as end devices of the access communications network that is connected with the public or private network. Because biometric data needs to be transmitted over a medium, identity authentication must consider not only the sophistication of the biometric method, but also the sophistication of an encryption technique, as well as an intelligent network that is able to detect unauthorized intervention.

In either case, the biometrics apparatus may be viewed as an end device of an access networks. The access network may be wireless (WiFi, GSM, 3G), wired (POTS, DSL) or fiber to the premises (FTTP). The security aspects for each access network type have been described in previous chapters and therefore the strengths and weaknesses of each type should be seriously considered when biometric data is transmitted.

12.5 BIOMETRICS REMOTE AUTHENTICATION

ID verification and access authorization is a comparison process of in situ extracted biometrics data with pre-stored data. The following text deals with pre-stored data at a remote database (Figure 12.1).

Remote databases are classified in:

A. *Centralized biometric databases.* The biometric apparatus extract features and transmits them across the network to a remote central database for comparison and verification. Results are sent back to the authenticating

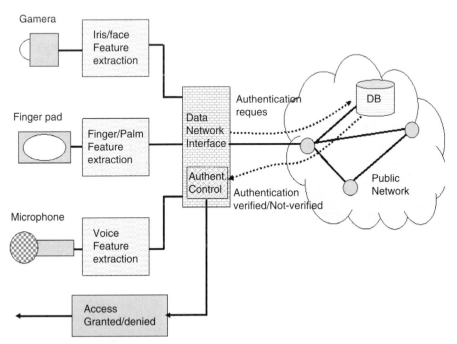

Figure 12.1. When the database is remote, biometric data are transmitted across the network over potentially vulnerable links.

apparatus over the network to grant or deny access. In this case, data are transmitted over potentially vulnerable links. Therefore biometrics data need to be encrypted, the links need to be secure and the database itself needs to be protected; the primary database should not be collocated with the protection database. However, both primary and protection databases need to be updated almost simultaneously.

B. *Distributed biometric databases.* The biometric apparatus maintains a col-located database, which periodically updates a regional database. Depending on database distribution, a centralized database may also be periodically updated by the regional databases for record keeping. In distributed data-bases, verification and authorization is fast (because of collocation) and does not risk unauthorized interception for each verification/authorization. However, updating the regional databases raises the same issues as the centralized database.

When transmitting biometric data over transmission links, a sophisticated intruder may capture data not for immediate use only but for use at a later time. Consequently, the sophistication of the authentication algorithm and the encryption method are both as important as is the security of the communications network and its medium. In addition, biometric data are unique to each person and in a manner of speaking they constitute a personal key with which authorized access to buildings, accounts, terminal and more is granted. To deter misuse of data, combined biometrics should be used in conjunction with strong encryption. Finally, the subject of biometrics and security has not been exhausted as research in this field continues.

REFERENCES

1. NIST Special Publication 800-36, "Guide to Selecting Information Technology Security Products".
2. USA Department of Defense, "*Department of Defense Biometric Standards Development Recommended Approach*", Alrington, Virginia, September 10, 2004.
3. FIPS PUB 201, Personal Identity Verification (PIV) of Federal Employees and Contractors, February 25, 2005.
4. NIST Special Publication 800-73 Supplementary Information, Namespace Management for Personal Identity Verification (PIV) Applications and Data Objects, October 6, 2005.
5. S.V. KARTALOPOULOS, "Associative RAM-based CAM Applicable to Packet-based Broadband Communications Systems", IEEE Globecom'98, Sydney, Australia, November 1998.
6. S.V. KARTALOPOULOS, "Communications Security: Biometrics over Communications Networks", Globecom 2006, San Francisco, on CD-ROM, NIS03-1, ISBN: 1-4244-0357-X, ISSN: 1930-529X.
7. S.V. KARTALOPOULOS, "Security and Biometrics in Modern Communications Networks", Plenary talk, 6th WSEAS Conference, 12/29–31/2007, Cairo, Egypt.

Acronyms

AAA = Authentication, Authorization, and Accounting

ADM = Add-Drop Multiplexer

ADSL = Asymmetric Digital Subscriber Line

AES = Advanced Encryption Standard

AH = Authentication Header

AID = Application Identifier

ALP = Application Layer Protocol

AODV = Ad-hoc On-demand Distance Vector routing

AON = All Optical Network

AP = Access Point

APDU = Application Protocol Data Unit; Authentic Protocol Data Unit

APS = Automatic Protection Switching

ARAN = Authenticated Routing for Ad-hoc Networks

ASE = Amplified Spontaneous Emission

ASK = Amplitude Shift Keying

ATM = Asynchronous Transfer Mode

AU = Administrative Unit

AU-n = Administrative Unit, level n

AuC = Authentication Center

B = B channel (ISDN)

BB = Broadband

BBO = Beta-Barium Borate

BER = Bit Error Rate; Basic Encoding Rules

B-ISDN = Broadband Integrated Services Digital Network

BIP-8 = Bit Interleaved Parity 8 field

bps = bits per second

BRI = Basic Rate Interface

BSC = Base Station Controller

BSS = Base Station System; Basic Service Set

BTS = Base Transceiver Station

BWA = Broadband Wireless Access

C-n = Container-level n; n = 11, 1, 2, 3, or 4

CAM = Content Addressable Memory

CAT = Chaos Annulling Trap

CBC = Cipher Block Chaining

CCMP = Counter mode with block Chaining Message authentication code Protocol

CCSDS = Consultative Committee for Space Data Systems

CDMA = Code Division Multiple Access

CFB = Cipher Feedback

CFT = Chaos Fixed Trap

CIT = Craft Interface Terminal

CMIP = Common Management Information Protocol

CMISE = Common Management Information Service Element

CMVP = Cryptographic Module Validation Program

CO = Central Office

CODEC = COder-DECoder

COP = Connection Oriented Protocol

CP = Customer Premises

CPE = Customer Premises Equipment

Security of Information and Communication Networks, by Stamatios V. Kartalopoulos
Copyright © 2009 Institute of Electrical and Electronics Engineers

CRC = Cyclic Redundancy Check

CRC-N = Cyclic Redundancy Check, width N

CSDC = Circuit Switched Digital Capability

CSMF = Conventional Single Mode Fiber

CSMD/CD = Carrier Sense Multiple Access/Collision Detection

CSP = Critical Security Parameters

CU = Channel Unit

CW = Continuous Wave

CWDM = Coarse Wavelength Division Multiplexer

D = D channel (ISDN)

DACI = Detection with Alarm and Countermeasure Intelligence

DACS = Detection with Alarm and Continuous Service

DADS = Detection with Alarm and Discontinuing Service

dB = Decibel

dBm = Decibel with 1 mWatt reference

DCC = Data Communication Channel

DCE = Data Circuit-terminating Equipment

DCF = Distributed Coordination Function

DCN = Data Communications Network

DCS = Digital Cross-connect System

DES = Data Encryption Standard

DFT = Discrete Fourier Transform

DH = Diffie-Hellman algorithm

DiffServ = Differentiated Services

DL = Data Link

DLC = Digital Loop Carrier

DLP = Discrete Logarithm Problem

DR = Dynamical Routing

DoS = Denial of Service

DS = Distribution System

DSAP = Destination Service Access Point

DSF = Dispersion Shifted Fiber

DSL = Digital Subscriber Line

DSLAM = Digital Subscriber Line Access Multiplexer

DS-n = Digital Signal level n; n = 0, 1, 2, 3

DSR = Dynamic Source Routing

DSS = Digital Signature Standard

DS-SMF = Dispersion Shifted Single Mode Fiber

DTE = Data Terminal Equipment

DWDM = Dense Wavelength Division Multiplexing

E0 = ITU-T G.703 Electrical interface signal 64 kbit/s

E1 = ITU-T G.703 Electrical interface signal 2048 kbit/s

E11 = ITU-T G.703 Electrical interface signal 1544 kbit/s

E12 = ITU-T G.703 Electrical interface signal 2048 kbit/s

EAP = Extensible Authentication Protocol

EAPOL = EAP Over LAN

ECB = Electronic Codebook

ECC = Elliptic Curve Cryptogaphy

ECDH = Elliptic Curve Diffie-Hellman

ECDLP = Elliptic Curve Discrete Logarithm Problem

ECDSA = Elliptic Curve Digital Signature Algorithm

ECMQV = Elliptic Curve Menezes-Qu-Vanstone

EDC = Error Detection Code

EDCV = Error Detection Code Violation

EDFA = Erbium-Doped Fiber Amplifier

E/O = Electrical to Optical

ESCON = Enterprise Systems Connectivity

ESP = Encapsulating Security Payload

FCAPS = Fault, Configuration, Accounting, Performance, and Security

FCC = Federal Communications Commission

FCS = Frame Check Sequence

FDM = Frequency Division Multiplexing

FDMA = Frequency Division Multiple Access

FEC = Forward Error Correction; Forward Equivalency Class

FFT = Fast Fourier Transforms

FIPS = Federal Information Processing Standard

FHSS = Frequency-Hopping Spread Spectrum

FITL = Fiber In The Loop

FM = Frequency Modulation; Fault Management

FOT = Fiber Optic Terminal

FOTS = Fiber Optic Transmission System

FPS = Fast Packet Switching

FSK = Frequency Shift Keying

FSO = Free Space Optical communications systems

FTTB = Fiber To The Building

FTTC = Fiber To The Curb

FTTCab = Fiber To The Cabinet

FTTD = Fiber To The Desk

FTTH = Fiber To The Home

FTTO = Fiber To The Office

FTTT = Fiber To The Town

FTTx = Fiber To The [any]

GbE = Gigabit Ethernet

Gbps = Gigabits per second

GBps = Gigabytes per second

Gbps = Gigabits per second = 1000 Mbps

gcd = greatest common divisor

GF = Galois Field

GFC = Generic Flow Control

GFP = Generic Framing Procedure

Ghz = Gigahertz (10^9 Hz)

GMSC = Gateway Mobile Services Switching Center

GIWU = GSM Interworking Unit

GSM = Global System for Mobile

HDLC = High level Data Link Control

HEC = Header Error Control

HMAC = keyed-Hash Message Authentication Code

HO = Higher Order

HOVC = Higher Order Virtual Container

HTTP = Hypertext Transfer Protocol

IBSS = Independent Basic Service Set

ICV = Integrity Check Value

ID = Identifier

IDI = Initial Domain Identifier

IDL = Interface Definition Language

IDLC = Integrated Digital Loop Carrier

IDSL = ISDN DSL

IE = Information Element

IEEE = Institute of Electrical and Electronics Engineers

IETF = Internet Engineering Task Force

IKMP = Internet Key Management Protocol

IMS = Information Management System; IP-based Multimedia Subsystem

IMSI = International Mobile Subscriber Identity

IN = Intelligent Network

I-NLSP = Integrated Network Layer Security Protocol

IntServ = Internet Services

IOF = Inter Office Framework

ION = Intelligent Optical Networks

IP = Internet Protocol

IPng = Internet Protocol next generation

IPSEC = Internet Protocol Security

IPv6 = Internet Protocol version 6

IPvn = Internet Protocol version n

ISAKMP = Internet Security Association and Key Management Protocol

ISDN = Integrated Services Digital Network

ISO = International Organization for Standardization

ISP = Internet Service Provider

ITL = Information Technology Laboratory

ITSP = Internet Telephony Service Provider

ITU = International Telecommunications Union

ITU-T = ITU Telecommunications
  Standardization Sector

IV = Initialization Vector

JCP = Java Community Process

Kbps = Kilobits per second = 1000 bps

KEES = Key Escrow Encryption System

KEK = Key Encryption Key

LAC = Link Access Control

LAN = Local Area Network

LASER = Light Amplification by
  Stimulated Emission or Radiation

LCAS = Link Capacity Adjustment Scheme

LD = Long Distance

LEC = Local Exchange Carrier

LED = Light Emitting Diode

LEOS = Low Earth Orbit Satellite

LH = Long Haul

LLC = Logical Link Control

LMDS = Local Multipoint Distribution
  Service

LO = Low Order

LPF = Low Pass Filter

LSB = Least Significant Bit

M1 = Level 1 Multiplexer

MAC = Message Authentication Code;
  Media-specific Access Control

MAN = Metropolitan Area Network

MANET = Mobile Ad Hoc Networks

MBps = Megabytes per second

Mbps = Megabits per second

MBps = Megabytes per second

Mbps = Megabits per second (1000 Kbps)

MD5 = Message Digest 5

ME = Mobile Equipment

MGCF = Media Gateway Control Function

MGW = Media Gateway

Mhz = Megaherzt (10^6 Hertz)

MIC = Message Integrity Code

MPDU = MAC Protocol Data Unit

MPI = Multiple Path Interference

MSB = Most Significant Bit

MSC = Mobile Services Switching Center

MSN = Mobile Service Node

msec = millisecond

μsec = microsecond

MUX = Multiplexer

mW = milliWatt

MXE = Message Center

NAP = Network Access Provider

NAS = Network Access Server

NASA = National Aeronautics and Space
  Administration (USA)

NE = Network Element

NEL = Network Element Layer

NF = Noise Figure

NGI = Next Generation Internet

NIC = Network Interface Card

NIST = National Institute of Standards and
  Testing

NIU = Network Interface Unit

NLSP = Network Layer Security Protocol

nm = nanometer

NMS = Network Management System

NNI = Network to Network Interface;
  Network Node Interface

NRM = Network Resource Management

NRL = Naval Research Laboratory (USA)

ns = nanosecond

NSA = National Security Agency

NSN = Network Service Node

NSP = Network Service Provider

NT = Network Termination

NTIS = National Technical Information
  Service

NTU = Network Termination Unit

OA = Optical Amplifier

OAM = Operations, Administration and
  Management

OADM = Optical ADM

OAMP = OAM and Provisioning Services

OC = Optical Carrier

OCG = Optical Channel Group; Optical Carrier Group

OCh = Optical Channel with full functionality

OC-n = Optical Carrier level n (n = 1, 3, 12, 48, 192)

ODL = Optical Data Link

ODU = Optical Data Unit

O-E = Optical to Electrical conversion

OEO = Optical-Electrical-Optical Converter

OFA = Optical Fiber Amplifier

OFDM = Orthogonal Frequency Division Modulation

OH = Overhead

OLSR = Optimized Link State Routing Protocol

OM = Optical Multiplexer

OMA = Open Mobile Alliance

OMU = Optical Multiplex Unix

ONU = Optical Network Unit

OOK = On-Off Keying

OPLL = Optical Phase-Locked Loop

OPU = Optical Payload Unit

OS = Operating System

OSA = Optical Spectrum Analyzer

OSC = Optical Supervisory Channel

OSI = Open System Interconnect

OSI-RM = Open System Interconnect Reference Model

OSNR = Optical Signal-to-Noise Ratio

OSS = Operations Support System

OTU = Optical Transport Unit

PAD = Packet Assembler and Disassembler

PAE = Port Access Entity

PCM = Pulse Coded Modulation

PDL = Polarization Dependent Loss

PDN = Packet Data Network; Passive Distribution System

PDU = Protocol Data Unit

PHY = Physical Layer

PIN = Personal Identification Number

PIV = Personal Identity Verification

PIX = Proprietary Extension

PKI = Public-Key infrastructure

PLL = Phase-Locked Loop

PMD = Polarization Mode Dispersion

PN = Prime Number

PON = Passive Optical Network

POP = Point Of Presence

POS = Packet Over SONET

POTS = Plain Old Telephone Service

PPP = Point-to-Point Protocol

PRNG = Pseudo-Random Number Generator

PS = Protection Switching

PSK = Phase Shift Keying; Pre-Shared Key

PSTN = Public Switched Telephone Network

PTE = Path Terminating Equipment

PVC = Permanent Virtual Circuit

PVP = Permanent Virtual Path

QC = Quantum Cryptography; Quantum Computers

QKD = Quantum Key Distribution

QM = Quantum Mechanics

QN = Quantum Network

QoS = Quality of Service; Quality of Signal

quBit = quantum Bit

quReg = quantum Register

RADIUS = Remote Authentication Dial-in User Service

RAND = Random Challenge

RAM = Random Access Memory

RAS = Remote Authentication Server

RC4/5/6 = Rivest Cipher version 4/5/6

RF = Radio Frequency

RFP = Request For Proposal

RID = Registered Identifier

Rijndael = Rijmen and Daemen

RIP = Routing Information Protocol

RM = Resource Management

RN = Random Number

RNG = Random Number Generator

ROM = Read Only Memory

RSA = Rivest, Shamir, Adleman

RSN = Robust Security Network

RSNA = RSN Associations

RT = Remote Terminal

RTT = Round Trip Time

SAID = Security Association ID

SAP = Service Access Point

SAR = Segmentation And Re-assembly

SCPS = Space Communication Protocol Standard

SCPS-FP = SCPS File Protocol

SCPS-NP = SCPS Network Protocol

SCPS-SP = SCPS Security Protocol

SCPS-TP = SCPS Transport Protocol

SDH = Synchronous Digital Hierarchy

SDNS = Secure Data Network

SDP = Session Description Protocol

SDSL = Symmetric DSL

SDU = Service Data Unit

SEAD = Secure Efficient Ad-hoc Distance vector routing protocol

SH = Short Haul

SHA-n = Secure Hash Algorithm n, n = 1, 256, 384, 512

SHS = Secure Hash Standard

SIG = Special Interest Group

SIM = Subscriber Identity Module

SIP = Session Initiation Protocol

SKIP = Simple Key-management for Internet Protocols

SL-N = Security Level N, N = 1, 2, 3 …

SLA = Service Level Agreement

SLC = Synchronous Line Carrier; Subscriber Loop Carrier

SLF = Subscriber Location Function

SMF = Single Mode Fiber

SMN = SONET/SDH Management Network

SMS = Short Message Service; SDH Management Sub-network

SN = Sequence Number

SNMP = Simple Network Management Protocol

SNP = Sequence Number Protection

SNR = Signal to Noise Ratio

SOHO = Small Office and Home

SONET = Synchronous Optical Network

SPM = Self Phase Modulation

SR = Short Reach

SRES = Signed Response

SS = Subscriber Station

SS7 = Signaling System #7

SSAP = Source Service Access Point

SSL/TLS = Secure Sockets Layer/Transport Layer Security

STA = Station

STM-n = Synchronous Transport Module level n (n = 1, 4, 16, 64)

STP = Shielded Twisted Pair

STS = Synchronous Transport Signal

SVC = Switched Virtual Circuit

T1 = A digital carrier facility used to transmit a DS1 signal at 1.544 Mbps

T3 = A digital carrier facility used to transmit a DS3 signal at 45 Mbps

TA = Terminal Adapter

TACACS = Terminal Access Controller Access Control System

Tbps = Terabits per second = 1000 Gbps

TBRPF = Topology Dissemination Based on Reverse-Path Forwarding

TC = Tandem Connection

TCP = Transmission Control Protocol

TCP/IP = Transmission Control Protocol/ Internet Protocol

TDEA = Triple Data Encryption Algorithm

TDM = Time Division Multiplexing

TDMA = Time Division Multiple Access

TE = Terminal Equipment

TEI = Terminal Endpoint Identifier

TEK = Traffic Encryption Key

THz = Tera-Hertz (1000 GHz)

TI = Trace Identifier

TIA = Telecommunications Industry Association

TINA = Telecommunications Information Networking Architecture consortium

TISPAN = Telecoms & Internet converged Services & Protocols for Advanced Networks

TKIP = Temporary Key Integrity Protocol

TL1 = Transport Language 1

TM = Traffic Management; Terminal Multiplexer

TMN = Telecommunications Management Network

TMSI = Temporary Mobile Subscriber Identity

TP = Twisted Pair

TTC = Tracking, Telemetry, and Command

TTP = Trusted Third Parties

TS = Time Stamp; Time Slot

TSI = Time Slot Interchanger

TTA = Telecommunications Technology Association

TU-n = Tributary Unit level n; n = 11, 12, 2, or 3

TUG-n = Tributary Unit Group n; n = 2 or 3

UDP = User Datagram Protocol

ULH = Ultra Long Haul

ULR = Ultra Long-Reach

UNI = User to Network Interface

USTIA = United States Telecommunications Industry Association

UTP = Unshielded Twisted Pair

UV = Ultra-violet

VC = Virtual Channel

VC-n = Virtual Container level n (n = 2, 3, 4, 11, or 12)

VC-n-Mc = Virtual Container level n, M concatenated Virtual Containers

VC-n-X = X concatenated Virtual Container-ns

VC-n-Xc = X Contiguously concatenated VC-ns

VC-n-Xv = X Virtually concatenated VC-ns

VCC = Connection

VCI = Virtual Circuit Identifier

VDSL = Very-high-bit rate DSL

VF = Voice Frequency

VLR = Visitor Location Register

VoIP = Voice over IP

VP = Virtual Path

VPC = VP Connection

VPI = Virtual Path Identifier

VPN = Virtual Private Network

VSA = Vendor-Specific Attributes

VSR = Very Short Reach

VT = Virtual Tributary

VoIP = Voice over IP; Video over IP

WDM = Wavelength Division Multiplexing

WEP = Wired Equivalent Privacy

WLAN = Wireless LAN

WMAN = Wireless Metropolitan Area Networks

WiMax = Worldwide Interoperability for Microwave Access

WPA = WiFi Protected Access

WPAN = Wireless Personal Area Network

WPON = WDM PON

WSN = Wireless Sensor Networks

WW II = World War II

xDSL = any-DSL

XOR = Exclusive Or

Index